NAVIGATING OFFSHORE GEOTECHNICAL CHALLENGES

A Compass to Problems Challenges and Solutions of Offshore Geotechnical Engineering

DR CR PARTHASARATHY

STARDOM BOOKS

www.StardomBooks.com

STARDOM BOOKS

112 Bordeaux Ct.

Coppell, TX 75019, USA

FIRST EDITION SEPTEMBER 2025

STARDOM BOOKS, LLC.
112 Bordeaux Ct. Coppell, TX 75019, USA

www.stardombooks.com

Stardom Books, United States
Stardom Alliance, India

NAVIGATING OFFSHORE GEOTECHNICAL CHALLENGES

A Compass to Problems Challenges and Solutions of Offshore Geotechnical Engineering

DR CR PARTHASARATHY

p. 290
cm. 13.5 X 21.5

Category: TEC026000: Technology & Engineering : Mining
TEC047000: Technology & Engineering : Petroleum
TEC009150: Technology & Engineering : Civil - Soil & Rock

ISBN: 978-1-957456-83-6

This book is dedicated to Lord Venkateshwara
&
To my beloved son CP Shreyas and
my wife, Mrs Anupama Parthasarathy, for their unwavering support throughout this journey.

Toni Morrison said,
"If there's a book that you want to read, but it hasn't been written yet, then you must write it"
And so, here it is - the book I wished existed, now written by me.

ACKNOWLEDGMENTS

I begin with heartfelt gratitude to the Almighty, my family, gurus, mentors, coaches, and to EVERYONE who ever said anything positive to me or taught me something. Without Divine Blessings, none of this would have been possible.

My deepest thanks to my wife, Mrs Anupama Parthasarathy, for her relentless encouragement and untiring support in both my professional and personal pursuits. To my son CP Shreyas, whose young and insightful mind continues to inspire new thoughts and ideas. Thank you, dears!

I sincerely thank my parents, Late C Ramanuja Iyengar & CR Leela, and my siblings, their children, the entire extended family on both sides, well-wishers and friends for their continuous support and encouragement.

I express my profound gratitude to Prof KS Subba Rao – Formerly Professor, Dept of Civil Engineering, Indian Institute of Science, Bangalore for his expert guidance, mentorship, and incessant encouragement throughout my professional career. It's been an honor to be under his guidance, and I feel humbled to be under his shadow.

With respect and gratitude, I thank Prof TG Sitharam, Chairman, All India Council of Technical Education, Govt of India, for his constant encouragement and continued support throughout my professional career and for writing the foreword for this book.

Sincere thanks to Prof VS Chandrasekaran, formerly professor, Department of Civil Engineering, Indian Institute of Technology, Bombay, for his consistent support and guidance.

I remain profoundly grateful to my previous employers, Mr. UM Raghavendra of Land & Offshore Foundations, Bangalore and Mr. Somehsa of Somehsa Geoscience Pte Ltd., Singapore, for providing opportunity and guiding me in my professional career in Onshore and Offshore Geotechnical engineering, respectively.

"Without the experiences and unwavering support from my peers and team at Sarathy Geotech & Engineering Services Pvt Ltd., (SGES), this book would not have been possible. I am deeply grateful for the opportunity to lead such an exceptional group-being a leader of great leaders is truly a blessed position.

Thank you, SGES-ians, for your dedication to the profession; some of the content in this book draws from your committed efforts on various projects. This book reflects your dedication, passion, and excellence. A special mention to Srinivas Nagappa, Satish Yarasi, and Kris Gustavian along with several other technicians for their invaluable contributions on the field.

I hereby acknowledge and convey sincere thanks to Dr Gangisetti Ramesh and Dr Reshma Sukumar for their tireless efforts in manuscript preparation, Sangeetha, Anusha Bekal, Nandhagopal, Sumanth Arun Talkad, Prashanth Talkad, Chanchal, Ashitha, Aravind, and Vijaya for initial manuscript & editorial assistance, and for making the book cover launch a success.

Sincere thanks to the team at Pile Dynamics Inc., USA for over two decades of technological support, knowledge sharing and mentorship - especially Dr Brent Robinson, Ryan Allins, Geoge Piscsalko and Anna Sellountou.

Gratitude to Er Mohammed Hussein, Partner – GRL Engineers and pile Dynamics Inc., USA, for insightful discussion, mentorship, and writing the foreword for this book.

Special thanks to Dr Prakasha, (Formerly CSD, Saudi Aramco; now, offshore Geotechnical consultant for every insightful discussion and for contributing a foreword. His unique perspective has reminded me that every problem has multiple solutions.

To all our customers, end clients, and project partners - thank you for the trust, opportunities, and constructive discussion during the course of several projects that have enriched this journey.

"Thanks to everyone on my publishing team – Raam and Ranjitha. Thanks to everyone at Scribe team for your unwavering support in bringing this book to life.

CONTENTS

PREFACE

This book, "Navigating Offshore Geotechnical Challenges" stems from a hope that has grown for nearly 25 years of immersion in the world of offshore engineering: the hope to provide a clear and reliable guide for those who venture into its intricate and demanding depths. My own journey into this field began in the relatively calm waters of Southeast Asia.

Here, in places like Malaysia, Thailand, Vietnam, Brunei and Indonesia, the soil beneath the seabed often presented a consistent and predictable layered structure, typically a normally consolidated profile extending to about 150 meters. The procedures for geotechnical investigation, the design of foundations, and the installation of piles, while requiring expertise, followed a straightforward path.

However, the Indian offshore, particularly the western region near Mumbai, threw me into a completely different reality. The soil here was a complex and unpredictable mix, a far cry from the uniformity I had encountered earlier. This stark contrast forced me to recalibrate my understanding of soil behavior and to seek entirely new approaches to foundation design and installation.

I found myself reaching out to experts across the industry, seasoned professionals within ONGC, professors at IITs, and contractors with prior experience in the region – piecing together knowledge from diverse sources to navigate this geotechnical labyrinth. This experience was a powerful lesson in the critical importance of continuous learning and adaptation in offshore engineering.

As my career progressed, I had the opportunity to extend my work to the Middle East. Here, in Abu Dhabi and Saudi Arabia, I encountered yet another set of unique subsurface conditions. To a certain extent, my Southeast Asian experience provided some helpful background, but largely, the soil stratigraphy and the associated challenges demanded entirely different solutions.

Once again, I embarked on a journey of discovery, delving into available literature, analyzing past data, and consulting with industry peers to understand the specific geotechnical risks and how to mitigate them effectively.

Through these diverse and challenging experiences across Southeast Asia, India, and the Middle East, a unifying realization began to solidify.

The hard-won knowledge, the practical wisdom gained from overcoming a wide array of geotechnical obstacles, could be immensely valuable to others in the field. Instead of spending years grappling with similar challenges and "reinventing the wheel," engineers, designers, and students could benefit from a resource that consolidates these lessons learned. This conviction became the driving force behind this book.

"Navigating Offshore Geotechnical Challenges " is designed to be more than just a collection of technical data and formulas. It is intended to be a practical guide, enriched by real-world case studies drawn from my work in Indonesia, Malaysia, India, Abu Dhabi, and Saudi Arabia. These case studies cover a spectrum of geotechnical issues encountered in offshore projects, including shallow foundation problems in marine environments, ground improvement techniques, challenges related to jack-up rig moves, mud mat installations, pipeline geotechnics and the complexities of pile installations, such as premature refusal.

Furthermore, recognizing the importance of industry perspectives, I have included insights from several industry leaders across the five geographical regions covered in the book. These leaders, including heads of oil companies, EPC contractors, and survey contractors, share their unique viewpoints on the geotechnical challenges they have faced in offshore installations, the strategies they have employed to overcome them, and their vision for the future of the industry. Their contributions add a valuable layer of context and practical relevance to the technical content.

Finally, I have also touched upon the significance of leadership and personal development in the offshore environment. Offshore work demands not only technical expertise but also resilience, discipline, mindfulness, and the ability to manage oneself effectively in demanding and often stressful situations.

While this book primarily focuses on geotechnical challenges, I have included a prelude to my upcoming book on leadership through personality development, offering some initial thoughts on these crucial aspects of an offshore professional's life. Perhaps these lessons is for all.

It is my sincere hope that "Navigating Offshore Geotechnical Challenges " will serve as a reliable and practical guide, providing direction and clarity as you navigate the ever evolving and complex world of offshore geotechnical engineering.

Dr CR Parthasarathy
Founder, Chairman & Managing Director
Sarathy Geotechnical & Engineering Services Private Limited.,
India | Indonesia | Vietnam | Abu Dhabi | Saudi Arabia
www.sarathygeotech.com
partha@sarathygeotech.com

FOREWORD

प्रो. टी. जी. सीताराम
अध्यक्ष
Prof. T. G. Sitharam
FNAE, DGE, FASCE, FICE (UK)
Ph.D. (Univ of Waterloo, Canada), D.Sc
Post Doc (Univ of Texas, @Austin USA)
Chairman

सत्यमेव जयते

अखिल भारतीय तकनीकी शिक्षा परिषद्
(भारत सरकार का एक सांविधिक निकाय)
(शिक्षा मंत्रालय, भारत सरकार)
नेल्सन मंडेला मार्ग, वसंत कुंज, नई दिल्ली-110070
दूरभाष : 011-26131498
ई-मेल : chairman@aicte-india.org
ALL INDIA COUNCIL FOR TECHNICAL EDUCATION
(A STATUTORY BODY OF THE GOVT. OF INDIA)
(Ministry of Education, Govt. of India)
Nelson Mandela Marg, Vasant Kunj, New Delhi-110070
Phone : 011-26131498
E-mail : chairman@aicte-india.org

Prof. T. G. SITHARAM
Chairman AICTE,
Professor, Department of Civil Engineering, Indian Institute of Science, Bangalore **(IISc)**
Former Director, Indian Institute of Technology, Guwahati.**(IITG).**

It is with great pleasure that I write the foreword for this timely and impactful book, Navigating Offshore Geotechnical Challenges, authored by Dr. CR Parthasarathy. This publication is a valuable contribution to the specialized field of offshore geotechnical engineering-a domain that demands both scientific precision and practical ingenuity.

Offshore infrastructure development is fraught with technical complexities, environmental uncertainties, and high-stakes decision making. In this context, Dr. Parthasarathy brings to the table a wealth of real-world experience, acquired over decades of active involvement in pioneering offshore projects across the globe. This book distills that experience into a well-structured, practitioner-oriented guide, making it both Informative and highly relevant.

A distinguished alumnus of the Indian Institute of Science. Bangalore (IISC) , and the founder of Sarathy Geotech & Engineering Services Pvt. Ltd. Dr. Parthasarathy has consistently demonstrated excellence in geotechnical consultancy. His firm's recognition through national and international awards further affirms his leadership in the field.

प्रो. टी. जी. सीताराम
अध्यक्ष
Prof. T. G. Sitharam
FNAE, DGE, FASCE, FICE (UK)
Ph.D. (Univ of Waterloo, Canada), D.Sc
Post Doc (Univ of Texas, @Austin USA)
Chairman

सत्यमेव जयते

अखिल भारतीय तकनीकी शिक्षा परिषद
(भारत सरकार का एक सांविधिक निकाय)
(शिक्षा मंत्रालय, भारत सरकार)
नेल्सन मंडेला मार्ग, वसंत कुंज, नई दिल्ली-110070
दूरभाष : 011-26131498
ई-मेल : chairman@aicte-india.org
ALL INDIA COUNCIL FOR TECHNICAL EDUCATION
(A STATUTORY BODY OF THE GOVT. OF INDIA)
(Ministry of Education, Govt. of India)
Nelson Mandela Marg, Vasant Kunj, New Delhi-110070
Phone : 011-26131498
E-mail : chairman@aicte-india.org

What sets this book apart is its ability to bridge academic knowledge with field-based applications. Covering critical areas such as site investigations, pile foundations, jack-up rig assessments, and pipeline geotechnics, the book offers a comprehensive view of offshore challenges along with tested methodologies and thoughtful solutions.

Dr. Parthasarathy's contributions go beyond this publication. His active engagement with professional bodies such as the Indian Geotechnical Society (IGS), the Indian National Academy of Engineering (INAE), and DFI-India reflects his commitment to the advancement of engineering practice and knowledge sharing I wholeheartedly commend Navigating Offshore Geotechnical Challenges to engineers, researchers, industry professionals, and students alike. It is not just a technical reference it is a compass for anyone navigating the demanding yet rewarding world of offshore geotechnics.

T.G. Sitharam
(Prof. T.G. Sitharam) 21/4/25

FOREWORD

Offshore installations are an essential part of the global infrastructure critical to the functioning of modern civilization. They are constructed around the world for a variety of purposes including petroleum and natural gas production, windmills for electrical energy generation, weather monitoring stations, navigational aids systems, and other applications and uses. Because of their location, offshore structures are designed and constructed to withstand the severe environmental conditions of open deep waters in order to provide safe and stable platforms for workers and machines. Offshore facilities are industrial wonders and engineering marvels.

Foundations for offshore structures present a wide variety of challenges rooted in difficulties in information obtainability, knowledge limitations, and inherent uncertainties at the planning, design, construction and implementation phases that often require extrapolations beyond available data and experience. Geotechnical engineering is the branch of civil engineering that employs scientific methods and experience to determine, evaluate, and apply the interrelationship between the geologic environment and engineered works. It is an integral and essential part of the work that contributes critical roles to the success of each project; and its broad nature is demonstrated by the large number of technical committees comprising the International Society for Soil Mechanics and Geotechnical Engineering.

A major source of risk on offshore projects is the uncertainty related to the portion of work associated with subsurface geotechnical conditions, such as for example deep foundations. A major role the geotechnical engineer plays is project-specific risk assessment and management, and related mitigation strategies by identifying and quantifying uncertainties for informed decision making at every stage of initial planning, site characterization and evaluation, assessment of alternate designs, construction, monitoring, and final delivery.

Probabilistic and deterministic methods are employed to approach risk assessment and management of geotechnical uncertainties.

Early pioneers of geotechnical engineering practice relied on "the observational approach" to develop an understanding of the behavior of earth materials under loads.

This approach has been continually enhanced by innovations in electronic instrumentation, field testing, scientific theoretical advancements, computer modelling and sophisticated numerical analyses techniques. These now make it possible to understand with greater accuracy the nonhomogeneous, nonlinear, anisotropic, and spatially variable nature and behavior of earth materials for application to engineered works, especially offshore structures.

With this book, Dr Parthasarathy adds a significant contribution to the engineering body of knowledge and general literature. It presents an outstanding comprehensive treatment of the complex subject of offshore geotechnical engineering by combining excellent technical expertise, extensive practical experience, and exceptional human perspective that Dr Parthasarathy distinctively possesses based on his many years of effective involvement and successful career in the offshore industry. In addition to the detailed academic and practical engineering content, valuable case studies from many regions around the world, the book also uniquely draws on the extensive knowledge, experience, expertise, and wisdom of many of the offshore industry's leaders in an engaging style that will appeal to engineers and general readers alike and benefit all on professional and personal levels.

Mohamad H. Hussein, P.E., L.M. American Society of Civil Engineers
Executive Director of Global Industry Relations
Pile Dynamics, Inc.
GRL Engineers, Inc.
Cleveland, Ohio USA

FOREWORD

Dear Dr CR Parthasarathy.

Navigating Offshore Geotechnical Challenges. This book encompasses and brings out geological variations in Asia, especially West and South Asia. It highlights how regional experience is as important as theoretical background.

This is a must read for those who want to pursue the field of offshore geotechnical engineering and a reference to look back throughout one's profession. Such a compendium of regional experiences is a rare find.

With regards,

Dr K Prakasha
Offshore Geotechnical Consultant
Bangalore

1
INTRODUCTION

1.1 The Offshore Realm

Our dependence on offshore energy and infrastructure is a defining characteristic of our time, driving engineering ever deeper into the planet's final frontier. The challenge: to harmonize the immense power of the ocean with the structures we build within it. This is the domain of offshore geotechnical engineering. This specialized field encompasses the design, construction, and long-term maintenance of foundations for critical marine structures—from the towering rigs that extract hydrocarbons to the elegant turbines that harness the wind. This discipline is not static; it evolves continuously, propelled by the need to overcome the formidable obstacles of the marine environment.

1.2 The Subsea Challenge

The offshore environment presents a daunting set of issues to the geotechnical engineer. We're confronted with geological conditions that defy easy categorization: unusual soil compositions, where pockets of gas lie hidden beneath the seabed, and where carbonate formations exhibit behaviors unlike anything encountered on land, the stratigraphies are orthogonally different in different geographies. Thus, these are the kinds of complexities that demand a deep understanding.

Then there are the relentless forces of nature: the crushing power of waves, the persistent push of currents, and the ever-present threat of wind. These elements combine to exert enormous loads on offshore structures, loads that fluctuate and cycle over decades. And beneath it all, the seabed itself can be a source of peril. Geohazards—submarine landslides, seismic activity, and the slow creep of geological change—require constant vigilance and sophisticated risk assessment.

The offshore geotechnical engineer works at the intersection of immense forces and subtle soil mechanics. It's a field where the cost of miscalculation can be enormous, where logistical challenges are inherent, and where structures are designed not for years, but for generations.

1.3 Foundations in a Hostile World

To meet these challenges, engineers have developed a range of foundation systems, each adapted to specific conditions:

- Steel Jacket Structures: The workhorse of the oil and gas industry, these structures mostly transmit their weight to the seabed through deep foundation (piles), installed deep into the seabed.
- Gravity-Based Structures: Massive structures that rest directly on the seabed, relying on their own weight for stability.
- Jack up Rigs (Mobile Offshore Drilling Unit – MODU): Mobile platforms that can elevate themselves above the waves, providing a stable platform for drilling.
- Mud mats: Temporary supports, essential for stabilizing subsea installations during the critical early stages of construction.
- Subsea Templates: Frameworks placed on the seabed to support a network of wells and related equipment.
- Wind Turbine Foundations: A rapidly evolving area, with designs ranging from monopiles to floating platforms, each with its own geotechnical demands.
- Pipelines: Serve as both transportation conduits and structural foundations, stabilized through burial, trenching, or anchoring to resist environmental forces and support subsea equipment, while adapting to varying seabed conditions and depths.

The choice of foundation is not a simple one. It's a decision that balances structural requirements, environmental conditions, and the characteristics of the seabed.

1.4 The Indispensable Survey: Procedures for Investigation (ISO 19901-8, 2014, ISSMGE 2005)

The success of any offshore project rests on a foundation of knowledge, and that knowledge comes from meticulous site surveys. Geophysical and geotechnical investigations are the essential tools that allow us to characterize the seabed, identify potential hazards, and design with confidence. A comprehensive survey is not merely a preliminary step; it's an ongoing process that informs every stage of a project, from initial planning to long-term monitoring. It's an investment in risk reduction, design optimization, and the enduring integrity of offshore structures.

An offshore site investigation unfolds in two complementary phases: geophysical and geotechnical surveys.

1.5 Geophysical Investigation: Seeing the Invisible

Imagine trying to map a hidden city. You couldn't just start digging; you'd want to use tools to give you a layout of the land first. That's the essence of geophysical surveying. These are the non-intrusive techniques that allow us to "see" beneath the seabed, to map the geological architecture and identify potential hazards before we ever disturb a grain of sediment. It's a bit like being a detective, using a range of scientific tools to piece together a subsurface puzzle. For the novice, it's a fascinating glimpse into how physics can be applied to understand the world around us. For the expert, it's a reminder of the power of these tools to provide critical insights in a challenging environment.

1.5.1 Methods

- **Bathymetry:** We begin by mapping the underwater terrain. Single-beam echo sounders give us a basic water depth profile, but for a truly detailed picture, we deploy multi-beam systems. These sophisticated tools sweep the seabed with a wide fan of sound waves, capturing thousands of data points with each pulse. The result is a stunningly detailed, three-dimensional model of the seabed, revealing hidden canyons, dramatic ridges, and subtle slopes that would otherwise remain invisible. This is fundamental information, crucial for everything from navigation to the placement of structures.
- **Side-Scan Sonar:** To understand the character of the seabed itself—is it smooth sand, jagged rock, or something in between?

We use side-scan sonar. This technology sends out high-frequency sound waves that bounce off the seafloor. The strength of the returning signal depends on the texture of the surface: a smooth bottom reflects sound waves differently than a rough one. In this way, we can create detailed images of the seabed, revealing not only its composition but also the presence of features like pipelines, shipwrecks, and even subtle variations in sediment. For the novice, this is like seeing the seabed come to life; for the expert, it's a valuable tool for assessing potential hazards and planning more detailed investigations.

- **Sub-Bottom Profiling (Pinger, Boomer and Sparker):** Sub-Bottom Profiling provides a more detailed look at the shallower layers of sediment. Using sound waves, this technique creates very high-resolution images of the uppermost part of the seabed, revealing fine-scale details like subtle changes in sediment type and the presence of small, buried channels or other features. This is particularly useful for understanding the near-surface conditions that can affect foundation design and installation.
- **Magnetometry:** The Earth's magnetic field can also be a valuable source of information. Magnetometers measure variations in this field, allowing us to detect the presence of metallic objects on or beneath the seabed. This is particularly useful for locating pipelines, cables, and unexploded ordnance—hidden hazards that could pose serious risks to offshore operations. For the novice, this is like using a giant metal detector; for the expert, it's a critical tool for ensuring safety and avoiding costly damage.

1.5.2 Materials

These investigations require a fleet of specialized vessels and a treasure trove of high-tech equipment. Survey vessels, often equipped with sophisticated dynamic positioning systems, hold their position with incredible accuracy, ensuring the precision of measurements. High-resolution echo sounders and side-scan sonar systems capture the intricate details of the seabed. Sub-bottom profilers provide detailed images of the shallow subsurface. Magnetometers measure the variations in the magnetic field and locate the buried pipelines, cables or hidden hazards.

A typical geophysical survey might include a range of specific tests, as described above, the following:

- Bathymetric mapping, to create detailed charts of the seafloor topography.
- Side-scan sonar imaging, to reveal the texture and features of the seabed.
- Sub-bottom profiling, to image the fine-scale stratigraphy of the shallow sediments.
- Magnetic surveys, to locate metallic objects.

But these are not just isolated pieces of information. They are carefully integrated and interpreted to create a comprehensive and coherent understanding of the seabed and the subsurface, guiding the subsequent geotechnical investigations and informing the critical decisions that must be made in the design and construction of offshore structures.

1.6 Geotechnical Investigation: Getting Down to Earth (or Seabed)

While geophysical surveys give us a big picture, geotechnical investigations involve getting our hands dirty—or rather, our equipment wet. This phase involves the direct sampling and testing of the seabed soils (may include in-situ testing viz., PCPT, Vane Shear Test, T-Bar and Ball Penetrometer test), providing the crucial quantitative data that engineers need to design foundations that will stand the test of time and the forces of nature.

If geophysical surveys are like that aerial flyover, giving you a sense of the city's layout, geotechnical investigations are like going down into the streets, examining the buildings up close, and even taking core samples of the materials they're made of.

It's a reminder that even with the most advanced technology, there's no substitute for direct observation and measurement.

1.6.1 Methods

- **Borehole Drilling:** At the heart of geotechnical investigation is borehole drilling. This is how we create the access points to the subsurface, allowing us to retrieve soil/rock samples and perform in-situ tests. A variety of drilling techniques are used, depending on the specific soil conditions and the depth we need to reach. Rotary drilling, using a rotating drill bit, is a common method, while wireline drilling and sampling allows us to retrieve core samples from much greater depths quicker, faster and with high quality. And coring techniques are essential for obtaining relatively undisturbed samples of the soil/rock, preserving its delicate structure and properties for detailed laboratory analysis.

- **Cone Penetration Testing (CPT):** Cone penetration testing (CPT) is another vital tool in our arsenal. This in-situ testing technique involves pushing an instrumented cone into the soil at a constant rate and measuring the resistance the cone encounters. This provides a continuous, high-resolution profile of the soil's properties with depth, including how hard it is to push through the soil (cone resistance), how much friction there is on the side of the cone (sleeve friction), and the excess pore water pressure in the soil. There are also variations on the basic CPT, such as CPTu, which specifically measures pore pressure, SCPTu, which adds a measurement of soil shear wave velocity (how fast shear waves travel through the soil), and dissipation tests, where we measure how quickly pore pressure dissipates after the cone is pushed in, which tells us something about how easily water flows through the soil (permeability). This is a bit like a high-tech probe sending back information about the soil's composition, in simple terms.
- **Vane Shear Testing (VST):** When we're working with cohesive soils, especially soft clays, we often use the vane shear test (VST) to measure their undrained shear strength. This involves inserting a four-bladed vane into the soil and then rotating it. The torque required to rotate the vane gives us a direct measure of the soil's resistance to shearing, which is a critical parameter for designing foundation.
- **Piston Sampling:** Although no samples retrieved from the ground are truly undisturbed, the degree of disturbance can be greatly minimized by using piston samplers. These specialized devices are designed to minimize any disturbance to the soil's internal structure during the sampling process, ensuring that the samples we bring back to the lab accurately reflect the conditions in the ground.

- **Gravity Coring:** For certain types of seabed sediments, we can use a simpler technique called gravity coring. This involves dropping a hollow tube from a vessel onto the seabed.

The tube penetrates the sediment under its own weight, collecting a core sample as it penetrates.

Geotechnical investigations require a range of specialized equipment; all deployed from specialized vessels **(Figure 1.6.1-1)**. Drill ships and jack-up platforms provide stable bases for drilling operations, while specialized CPT deployment systems, often using powerful hydraulic systems, are used to push the cone into the seabed.

Piston samplers and gravity cores are used to collect soil samples. Where cemented layers are encountered, coring techniques are utilized to retrieve high quality core samples. Once those samples are retrieved, a whole suite of sophisticated laboratory equipment is needed to perform the tests that will reveal their properties.

It's a reminder of the complex logistics and advanced technology that are essential for working in the offshore environment.

Figure 1.6.1-1: Dedicated Geotechnical Survey Vessel
(Courtesy: Safeen Subsea survey, UAE)

1.6.2 Prognosis

Geotechnical investigations are carried out to provide the essential data that we need to:

- **Determine the soil stratigraphy:** to identify the different layers of soil that make up the seabed and to describe their characteristics.
- **Measure key soil properties:** including the soil's strength (how much load it can support), stiffness (how much it will deform under load), compressibility (how much it will undergo volume change), and permeability (how easily water flows through it).
- **Assess immediate settlement and consolidation potential:** to predict how much the soil will settle or compress over time (volume change) under the weight of a structure.
- **Evaluate foundation suitability:** to determine whether the soil conditions at a particular location are suitable for a given type of foundation, and if not, what modifications or alternative designs might be needed.
- **Ensure Quality:** The quality of the data we obtain from geotechnical investigations is critical. It's the foundation upon which our designs are built, and it directly impacts on the safety and long-term performance of offshore structures. To ensure we get the best possible data, we adhere to rigorous quality control procedures at every stage of the investigation. These include:
  - **Proper handling and preservation of soil samples:** to minimize disturbance and maintain their in-situ properties.
  - **Accurate execution of in-situ testing procedures:** with careful attention to equipment calibration and adherence to strict operational protocols.
  - **Adherence to rigorous laboratory testing protocols:** following established standards and best practices.
  - **Comparison of results from different testing methods:** to identify any inconsistencies or anomalies and to ensure the reliability of our data.
  - **Strict adherence to relevant industry standards and guidelines.**

- **Skilled and trained resources must be the ONLY personnels allowed to do the tests.**

1.6.3 Tests Performed

A wide range of geotechnical tests are performed on the soil samples we collect, both in the field and in the laboratory. These tests are designed to measure the key properties of the soil and to characterize its behavior under various conditions. Typical tests include:

- **Soil classification tests:** such as grain size analysis and Atterberg limits, which help us to determine the basic type of soil we're dealing with.
- **Index property tests:** such as water content, unit weight, and specific gravity, which provide further information about the soil's physical state.
- **Strength tests:** such as triaxial tests, direct shear tests, Torvane, Miniature vane, pocket penetrometer and unconfined compression tests, which measure the soil's undrained shear strength.
- **Consolidation tests:** such as oedometer tests, which determine the soil's compressibility and volume change over time.
- **Permeability tests:** such as falling head tests and constant head tests, which measure how easily water flows through the soil.
- **Cyclic loading tests:** which simulate the effects of repeated loading from waves and wind, helping us to understand how the soil will respond to these dynamic forces.

Dedicated geotechnical drilling vessels are normally equipped with a basic soil testing laboratory. This will facilitate the extrusion and description of samples, sub- sampling of specimens for immediate testing and waterproof re-sealing of the remaining sample. As a minimum and good practice, for every meter of samples retrieved, testing would typically include the determination of:

- Moisture content
- Wet and dry density
- Undrained shear strength, by means of:
  - Pocket penetrometer
  - Torvane
  - Miniature shear vane
  - Unconsolidated, undrained (UU) triaxial compression test.

- Reaction to Hydrochloric acid (HCl)

Effective re-sealing of samples for transport to the onshore laboratory is typically achieved by means of:

- Removing surface moisture/drilling mud
- Wrapping in clear plastic wrap ("cling film")
- Overwrapping in aluminum foil
- Placing in a cylindrical cardboard carton which is subsequently filled with molten paraffin wax, which, when cooled and hardened, provides a very effective water-tight seal.

1.6.4 Outcome

Geotechnical investigations culminate in a set of key deliverables that provide a comprehensive record of the site conditions. These typically include:

- **Borehole logs:** which provide a detailed description of the soil layers encountered in each borehole, along with information about the samples that were recovered.
- **CPT profiles:** which show how the measurands viz., cone resistance, sleeve friction, and pore pressure vary with depth, giving us a continuous picture of the soil's properties.
- **Laboratory test reports:** which present the detailed results of all the laboratory tests that were performed on the soil samples.
- **Geotechnical data reports:** which synthesize all the findings from the investigation, provide a comprehensive interpretation of the site conditions and offer recommendations for foundation design and construction.

Ultimately, it is the integration of the findings from both geophysical and geotechnical investigations that allows engineers to develop a truly complete and reliable picture of the subsea environment. This integrated understanding is essential for designing offshore foundations that are not only safe and reliable but also cost-effective and perfectly suited to the specific challenges of the site. It's a glimpse into the fascinating world of soil mechanics and the importance of careful investigation - a reminder of the need for continuous learning and the application of best practices in a field where precision and accuracy are paramount.

1.7 A Journey Across Diverse Geographies

Beyond this introduction, we embark on a detailed exploration of offshore geotechnical engineering across five key regions. We will delve into the unique geological settings of Offshore Malayasia, Indonesia, East Coast & West coast India, Abu Dhabi and Saudi Arabia. In each region, we will examine the characteristic soil conditions, the challenges they present, and the engineering solutions they demand, through case studies and in-depth analysis. Further chapters of this book will illustrate how the principles of offshore geotechnical engineering are applied in practice, and how they are constantly being refined to meet the evolving needs of the offshore industry.

PS: All the descriptions in this chapter and the subsequent chapters are from the author's personal experience.

2
INDIA

2.1 Introduction

As described by Magoon et al (1994), the key components of a petroleum system consist of the petroleum source rock, reservoir rock, seal rock, and overburden rock. Processes such as trap formation, hydrocarbon generation, migration, and accumulation are vital for boosting prospectively. It is essential to ensure the proper chronological arrangement of these elements alongside the processes to facilitate the accumulation and preservation of hydrocarbons. Exploration places and prospects are typically identified in basins or regions where there is a reasonable probability of a complete petroleum system being present. As a peninsula, India boasts abundant oil and natural gas resources. These reserves are identified across both onshore and offshore environments, with the offshore reserves further categorized into the West Coast and East Coast reserves. Notable offshore fields, either operational or in various stages of exploration, include:

Major Fields on the West Coast of India

1. **Bombay High Field:** Located in the Arabian Sea off the coast of Mumbai, Maharashtra. It is one of the largest offshore oil and gas fields in India.
2. **Daman Field:** Located in the Tapti-Daman block in the Arabian Sea. It primarily produces natural gas.
3. **Panna-Mukta-Tapti (PMT) Fields:** Located in the Arabian Sea, jointly owned by ONGC, Reliance Industries Limited (RIL), and BG Group. They produce both oil and gas.
4. **Ratna Field:** Situated offshore from the coast of Gujarat in the Arabian Sea. It primarily produces oil.

Major Fields in the East Coast of India

1. **Krishna-Godavari Basin (KG Basin)**: This basin is one of India's most prolific deep water hydrocarbon basins, located off the east coast. It includes several fields, including the KG-D6 block operated by Reliance Industries Limited, which has been a major contributor to India's gas production.
2. **Ravva Field**: Situated off the coast of Andhra Pradesh in the Krishna-Godavari Basin. It produces both oil and gas.
3. **Mahanadi Basin**: Located off the coast of Odisha, the Mahanadi Basin is another significant area for oil and gas exploration. It includes fields such as the Dhirubhai-1 and 3 (D1 and D3) gas fields operated by Reliance Industries Limited.
4. **Cauvery Basin**: Located off the coast of Tamil Nadu, the Cauvery Basin is known for its oil and gas potential. This basin includes the PY-3 field operated by the Oil and Natural Gas Corporation (ONGC).
5. **Andaman Basin**: Located in the Bay of Bengal, the Andaman Basin holds deep water exploration potential for oil and gas. However, commercial production from this basin has yet to be realized.

As of April 1st, 2022, India's estimated crude oil reserves amounted to 651.77 million metric tons (MMT), compared to 591.92 MMT the previous year. Geographically, most of these reserves, accounting for 33%, are in the Western Offshore region. And the estimated reserves of natural gas stood at 1138.67 billion cubic meters (BCM). The Western Offshore region holds the highest reserves of natural gas at 29.6%, followed by the Eastern Offshore region at 23.6%. (Ministry of Statistics and Programme Implementation 2023).

A comprehensive understanding of each offshore location's geology and geotechnical properties is imperative for the seamless execution of the upstream-downstream cycle in oil and gas production. This chapter provides an overview of the geology and stratigraphy of both the West and East Coasts of India. Additionally, it examines potential geotechnical risks associated with the installation of jacket platforms, positioning of jack-up rigs, laying of pipelines, etc. A few case studies are included to illustrate these challenges.

2.2 General Geology of the Region

Figure 2.2-1 shows different Oil & Gas reserve basins.

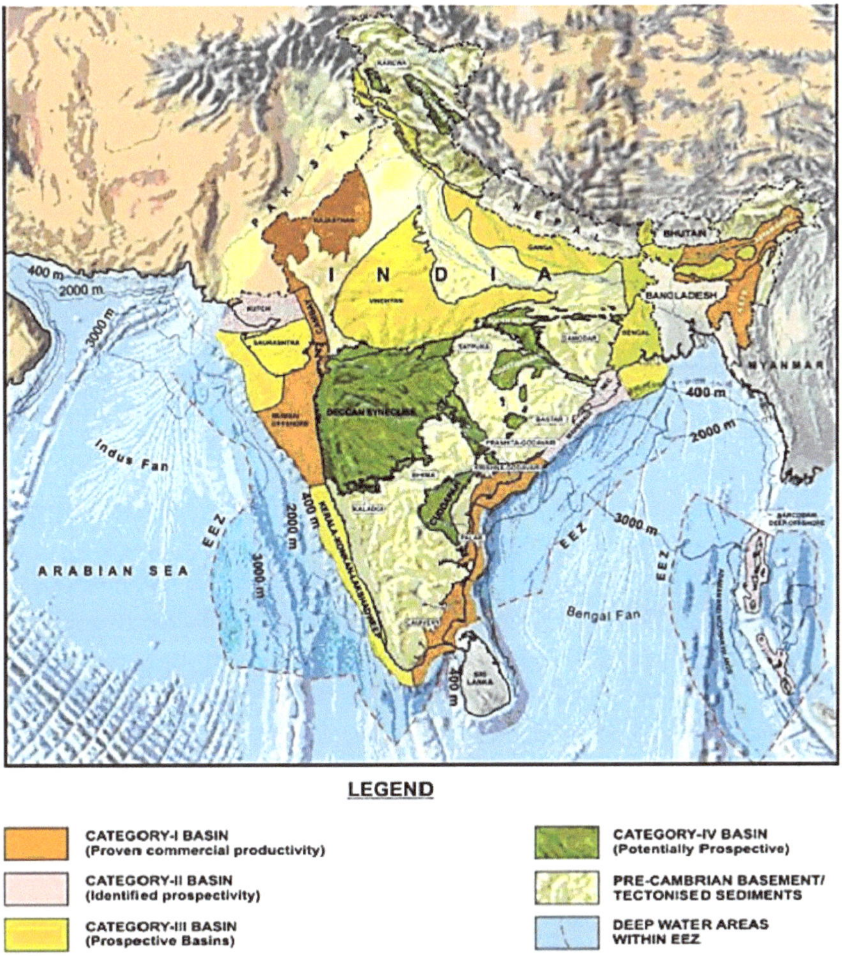

Figure 2.2-1: Basins of India (Directorate General of Hydrocarbons, Government of India. The Kerala-Konkan Basin. Accessed 9th July 2025) https://dghindia.gov.in//assets/downloads/56cfd920029d9The_Kerala-Konkan_basin.pdf.

2.2.1 West Coast Region

The West Coast of Mainland India (WCMI) exhibits a wide continental shelf, stretching from a few kilometers off the southern coast of Cochin to over 300 km off the Saurashtra-Mumbai coastline in the north. The current configuration of the WCMI has resulted from a sequence of tectonic events, including episodes of rifts over time, sedimentation processes, and erosion.

The western offshore area of India is subdivided into five pericratonic basins, namely the Kutch offshore, Saurashtra offshore, Bombay offshore, Konkan offshore, and Kerala offshore basins, arranged from north to south (Sriram et al. 2006). These basins extend from the continental shelf into deeper waters and are marked by east-west trending positive basement features, including the Saurashtra Arch, Son-Narmada Lineament, Vengurla Arch, and Tellicherry Arch. These geological features have played a crucial role in shaping the evolution of the basins. Among them, the Saurashtra Arch stands out due to its considerable size and the intensity of its tectonic activity. Serving as a significant boundary, it separates the northern Kutch basin from the southern Saurashtra basin, extending from the landmass across the shelf and into the depths of the ocean.

2.2.2 East Coast Region

The East Coast of India is represented by the rifted passive margin that had evolved in response to the process of continental rifting and subsequent seafloor spreading between India, Antarctica, and Australia during early Cretaceous breakup of Eastern Gondwanaland (Powell et al., 1988). The eastern Indian shelf spans approximately 2000 km in length, as a slender, elongated belt from West Bengal basin in the north to Tamil Nadu in the south. This extensive coastline comprises five major sedimentary basins: the Cauvery, Palar, Krishna–Godavari (K–G), Mahanadi, and Bengal basins.

The Mahanadi basin and the Krishna-Godavari basin are separated by a prominent fault line called the Vishakhapatnam high. Through extensive geological studies conducted by organizations such as ONGC (Oil Natural Gas Corporation of India) and GSI (Geological Survey of India), surficial sedimentary rock types within the East Coast basins, as well as the adjoining crystalline rocks, have been mapped. Apart from the east coast basins and their adjacent offshore areas, there is the Andaman offshore basin situated in the southeastern corner of Indian territory. This deepwater basin lies along the Andaman trench–arc region, serving as an active plate boundary between the Indian and Eurasian plates.

The offshore Cauvery basin extends along the south-eastern coast from Ramnath Puram near the Palk Strait to the area north of Madras along the Coromandel Coast encompassing an offshore area of about 95,000km^2.

2.3 General Stratigraphy of Offshore India

The following section provides a basic understanding of the typical generalized soil stratigraphy of some of the most important offshore oil and gas fields in India **(Tables 2.3.1-1, 2.3.1-2, 2.3.1-3)**

2.3.1 West Coast

Mumbai High Field

Depth (m)	Description	Shear Strength, S_u (kPa)		Angle of Internal Friction (ϕ), deg
		(Top)	(Bottom)	
0 – 11	Very Soft to Soft Calcareous Silty CLAY	2	10	-
11 – 15	Firm Calcareous Silty CLAY	20	70	-
15 – 20	Very Stiff Calcareous Silty CLAY	100	150	-
20 – 35	Very Stiff Calcareous Silty CLAY	100	200	-
35 – 45	Weakly to Moderately Cemented Siliceous Carbonate SAND	-	-	32
45 – 51	Very Stiff Calcareous Silty CLAY	80	200	-
51 – 70	Very Stiff Calcareous Silty CLAY with occasional shell fragments	125	220	-
70 – 75	Well-Cemented Siliceous Carbonate SAND	-	-	35
75 – 80	Very Stiff Calcareous Silty CLAY with occasional shell fragments	150	200	-
80 - 85	Dense Siliceous Carbonate SAND			33
85 – 90	Very Stiff Calcareous Silty CLAY with partings of fine sand	120	150	-
90 – 92	Well-Cemented Siliceous Carbonate SAND	-	-	35
92 – 95	Very Stiff Calcareous Silty Sandy CLAY with shell fragments	120	150	-
95 – 100	Well-Cemented Siliceous Carbonate SAND	-	-	35
100 – 125	Very Stiff Calcareous Silty CLAY with thin layers of well-cemented sand at multiple depths	150	250	-

Table 2.3-1-1: Generalized Soil Stratigraphy of Mumbai High Field

Bassein Field

Depth (m)	Description	Shear Strength, Su (kPa)		Angle of Internal Friction (ϕ), deg
		(Top)	(Bottom)	
0 – 3.5	Very Dense Clayey Silty SAND (Slightly cemented)	-	-	33
3.5 – 6	Very Stiff Calcareous Silty Sandy CLAY	120	175	-
6 – 20	Stiff to Very Stiff Calcareous Silty Sandy CLAY	160	75	-
20 – 25	Very Dense Siliceous Carbonate SAND with thin cemented layers	-	-	35
25 – 30	Very stiff Carbonate Silty Sandy CLAY	175	150	-
30 – 35	Weakly to moderately Cemented Siliceous Carbonate SAND	-	-	33
35 – 38	Very Stiff CLAY	150	180	-
38 – 43	Hard Clayey SILT with occasional shell fragments	250	350	-
43 – 46	Dense Silica SAND with lot of shell fragments	-	-	34
46 – 70	Very Stiff Calcareous Silty CLAY	220	150	-
70 – 75	Well-Cemented Siliceous Carbonate SAND with numerous shell fragments	-	-	35
75 – 100	Very Stiff Carbonate Silty Sandy CLAY with thin cemented sand layers & sand pockets	200	170	-
100 – 105	Moderately to Well-Cemented Siliceous Carbonate SAND with few shell fragments	-	-	34
105 – 120	Very Stiff to Hard Carbonate Silty Sandy CLAY	200	220	

Table 2.3.1-2: Generalized Soil Stratigraphy of Bassein Field

North Tapti

Depth (m)	Description	Shear Strength, Su (kPa)		Angle of Internal Friction (ϕ), deg
		(Top)	(Bottom)	
0 – 3	Loose Siliceous Carbonate SAND	-	-	25
3 – 11	Medium Dense Siliceous Carbonate SAND with thin Cemented Sand layers	-	-	32
11 – 13	Stiff Carbonate CLAY	75	100	-
13 – 20	Stiff to Very Stiff Carbonate Silty CLAY	75	175	-
20 – 25	Very Stiff Calcareous CLAY, with few pockets of Silt towards bottom, with Thin cemented Sand layers	100	160	-
25 – 35	Very Stiff Carbonate Silty CLAY, Highly Plastic, with Thin cemented Sand layers	120	150	-
35 – 37	Very Stiff SILT	80	100	
37 – 45	Hard Carbonate Silty CLAY	200	250	-
45 – 65	Hard Carbonate Silty CLAY, with numerous pockets of Silt and Sand, occasional Shell fragments at bottom	300	450	-
65 – 80	Very Hard Carbonate CLAY	350	500	-
80 – 85	Very Hard Siliceous Carbonate Clayey SILT	450	500	-
85 – 122	Very Hard Carbonate Silty CLAY with numerous pockets of Silt and few shell fragments and thin cemented sand layers	300	600	-
122 –125	Very Hard Carbonate Silty CLAY with numerous pockets of Silt and Fine Sand and occasional shell fragments	500	700	-

Table 2.3.1-3: Generalized Soil Stratigraphy of North Tapti Field

2.3.2 East Coast (Tables 2.3.2-1 and 2.3.2-2)

Digha Field

Depth (m)	Description	Shear Strength, Su, (kPa)		Angle of Internal Friction (ϕ), deg
		(Top)	(Bottom)	
0 – 3	Very Soft Sandy CLAY with seams of Fine Sand and traces of Shell Fragments	3	10	-
3 – 8	Loose Silty Fine SAND with traces of Shell Fragments and pockets of Clay	-	-	30
8 – 15	Medium Dense Silty Fine SAND with little Shell Fragments and traces of Clay pockets	-	-	32
15 – 22	Stiff Silty CLAY with some partings of Silt and few seams of Sand, traces of Shell Fragments	70	110	-
22 – 45	Medium to Dense Silty Fine SAND with few Shell Fragments and traces of Clay pockets	-	-	33
45 – 50	Very Stiff Silty CLAY with few pockets of Silt and traces of Shell Fragments	120	100	-
50 – 60	Very Stiff Clayey SILT interlayered with pockets of clay, sand and Ferrous and some seams of sand	125	225	-

Table 2.3.2-1: Generalized Soil Stratigraphy of Digha Field

Krishna-Godavari Basin

Depth (m)	Description	Shear Strength, Su, (kPa)		Angle of Internal Friction (ϕ), deg
		(Top)	(Bottom)	
0 – 6.9	Very soft to soft CLAY (Under Consolidated)	1	11	-
6.9 – 21	Very soft to firm CLAY (Under consolidated)	11	15	-
21 – 53.4	Soft to Stiff CLAY (under Consolidated)	15	52	-
53.4 – 90	Firm to Stiff CLAY (Under Consolidated)	52	40	-
90 – 97	Firm to Stiff CLAY (Under Consolidated)	40	50	-
97 – 118	Firm to Stiff CLAY (Under Consolidated)	50	54	-
118 – 147	Stiff to Very stiff CLAY (Under Consolidated)	54	200	-
147 – 160	Medium dense silty SAND	-	-	32

Table 2.3.2-2: Generalized Soil Stratigraphy of Krishna-Godavari Basin Field

2.4 Geological and Geotechnical Aspects of Offshore East Coast India

The offshore region along the East Coast of India, extending from Tamil Nadu to Odisha, represents a complex interplay of geological history, sedimentary dynamics, and geotechnical challenges (Schematic representation in **Figure 2.4-1**.

The subsequent sections provide a comprehensive analysis of the key geological and geotechnical characteristics of this region, emphasizing their implications for offshore engineering activities such as hydrocarbon exploration, renewable energy projects, and infrastructure development.

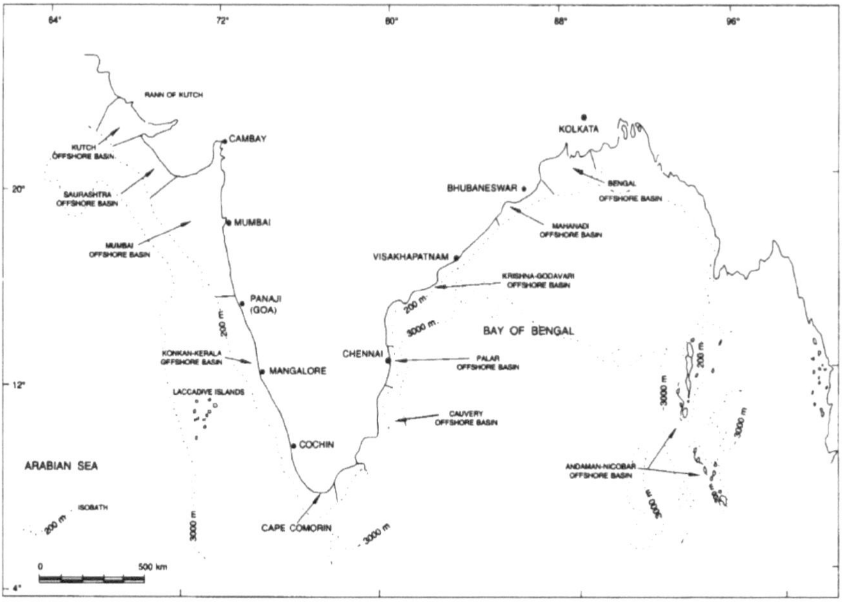

Figure 2.4-1: Offshore Basins of India

2.4.1 Geological Framework Basin Evolution and Tectonics

The Eastern Continental Margin of India (ECMI) formed during the breakup of Gondwana (~130 million years ago), creating rift basins such as Krishna-Godavari (KG), Mahanadi, Cauvery, and Palar. These basins evolved through distinct phases:

- Northern ECMI (north of 16°N): Dominated by rift structures with horst-graben systems.
- Southern ECMI: Influenced by shear forces during continental separation, transitioning from pull-apart to rift basins.

Seismic and gravity data reveal sedimentary thickness exceeding 10,000 m in deep offshore regions, with Mesozoic-Cenozoic sequences hosting hydrocarbon-bearing strata. The KG Basin alone holds proven reserves of over 40 trillion cubic feet (Tcf) of hydrocarbons, driven by Cretaceous-Paleogene source rocks and deltaic reservoir systems (Bastia and Radhakrishna, 2012), Directorate General of Hydrocarbons (2024).

2.4.2 Sediment Provenance and Deposition

Major rivers like the Mahanadi, Krishna-Godavari, and Cauvery deliver fine-grained sediments (clays, silts) enriched in quartz, feldspar, and clay minerals (montmorillonite, chlorite, kaolinite).

Geochemical signatures indicate high carbonate content (23%) and organic matter (7%) in marine clays, influencing soil plasticity and cohesion.

Shallow gas chimneys and mud volcanoes linked to vertical migration pathways are common in these basins.

2.4.3 Deltaic Systems and Sedimentary Dynamics

The region is dominated by sedimentary processes influenced by major east-flowing rivers such as the Mahanadi, Godavari, Krishna, and Cauvery. These rivers deposit vast amounts of sediments offshore, forming deltaic systems and estuaries.

The Krishna-Godavari Basin features coastal, deltaic, shelf-fan, deep-sea channel systems, and deep-water fan complexes. Continuous formation of sand bars and spits has altered river outlets over time, with northerly shifts observed due to Coriolis forces and sediment deficit in areas like Subarnarekha to Godavari.

2.4.4 Shallow Gas and Mud Volcanoes

Shallow gas chimneys and mud volcanoes are linked to vertical migration pathways, particularly in the KG Basin. These features impact geotechnical assessments for offshore foundations and pipelines.

They also pose risks of gas escape and pressure-induced instability during drilling operations.

2.4.5 Mass Transport Complexes (MTCs)

MTCs, covering 500–1,000 km² in the KG Basin offshore, create uneven seabed that complicate pipeline routing and infrastructure placement. Buried MTCs necessitate detailed geophysical surveys to ensure safe infrastructure placement.

2.4.6 Slope Instability

Rapid sedimentation rates (2–5 m/kyr in the KG Basin), seismic shocks, and gas hydrate dissociation contribute to slope instability (Ramprasad et al. 2011). Hotspots include steep slopes (>15°) off Chennai and Krishna-Godavari canyons. Mitigation strategies involve slope angle monitoring (<5° for stability) and subsea berms to prevent failures.

2.4.7 Geotechnical Characteristics

Soil Composition

The offshore East Coast is primarily composed of fine-grained sediments such as clays and silts deposited by river systems. Nearshore sediments are typically less consolidated due to active deposition processes, while deeper sediments exhibit varying degrees of consolidation depending on their depth and proximity to the coastline. Marine clays often exhibit high plasticity indices (PI > 40%) and low permeability ($<10^{-9}$ m/s), and under consolidation behavior leading to prolonged pore pressure dissipation after installation of foundations.

Shear Strength Parameters

Undrained shear strengths range between 10–50 kPa generally in the top 50m and increases further with depth. Sands show drained peak friction angles of 30°–38°, depending on grain size distribution and history of sedimentation levels. Coarse-grained sands are interbedded with clays at varying depths, posing heterogeneity challenges.

Consolidation Behavior

Due to the high compressibility of marine clays, time-dependent settlements can be significant. Pre-consolidation pressures vary widely, reflecting differences in depositional history. It is observed that the cohesive deposits exhibits under-consolidation with OCR < 0.5 in this region. Engineers must account for secondary compression effects in long-term designs.

Permeability and Pore Pressure

Marine clays exhibit extremely low permeability ($<10^{-9}$ m/s), leading to prolonged pore pressure dissipation after installation of foundations. Sands, conversely, allow rapid drainage, reducing risks of excess pore pressures.

Liquefaction Potential

Loose saturated sands in deeper waters pose a risk of liquefaction during earthquakes. Detailed site-specific analyses using cyclic stress ratios and in-situ cone penetration tests are essential to assess vulnerability to liquefaction.

2.5 Key Geohazards and Mitigation Strategies

2.5.1 Slope Instability

Rapid sedimentation rates (2–5 m/kyr in the KG Basin), seismic shocks, and gas hydrate dissociation contribute to slope instability. Hotspots include steep slopes (>15°) off Chennai and Krishna-Godavari canyons.

Mitigation strategies involve slope angle monitoring (<5° for stability) and subsea berms to prevent failures (Ramprasad et al. 2011).

2.5.2 Mass Transport Complexes (MTCs)

MTCs in the KG Basin offshore create uneven seabed that challenge pipeline routing. Buried MTCs necessitate detailed geophysical surveys to ensure safe infrastructure placement (Ramprasad et al. 2011).

2.5.3 Shallow Faults

Post-Miocene faults with displacement <50m are detectable via 3D seismic data. These faults influence foundation design and require careful assessment to ensure structural integrity.

2.5.4 Cyclonic Activity and Environmental Considerations

Frequent cyclones in the Bay of Bengal pose risks to offshore structures. Constant changes in sediment deposition patterns demand adaptive engineering designs. Preservation of estuarine ecosystems is critical during offshore development, emphasizing the need for sustainable practices.

2.6 Geological Framework of Offshore West Coast India

2.6.1 Tectonic Evolution

The western offshore basins (Mumbai Offshore and Konkan-Kerala) formed during the Late Cretaceous due to rifting associated with the breakup of Madagascar (~90–84 Ma) and Seychelles (~65 Ma) from India. These basins exhibit horst-graben structures bounded by NW-SE to N-S trending basement faults, creating segmented shelf margins. The Mumbai Offshore Basin spans ~230 km in width, while the Konkan-Kerala Basin averages 75 km, transitioning into deepwater ridges like the Laxmi-Laccadive Ridge (Singh, Radhakrishna, and Pant 2007). The region's tectonic framework was further influenced by the Carlsberg Ridge, an active spreading center in the Arabian Sea, which drives ongoing subsidence and sediment deposition (Jain, Banerjee, and Kale 2020)

2.6.2 Stratigraphy

Deccan Traps

Late Cretaceous flood basalts (~65 Ma) underlie Tertiary sediments, with volcaniclastics influencing near-surface geotechnical properties. These basalts originated from fissure-type eruptions along transform faults during the Deccan volcanism event (Schoene et al. 2015).

Syn-Rift Sediments

Paleocene-Eocene clastic and carbonate sequences (e.g., Panna, Bassein formations) form hydrocarbon reservoirs. These deposits include fluvial-lacustrine sediments and carbonate platforms, with source rocks matured by thermal subsidence (Deshmukh et al. 2017).

Post-Rift Deposits

Oligocene-Miocene deltaic and marine sediments (e.g., Daman Formation) dominate the outer shelf, with thickness exceeding 3 km in grabens. These deposits consist of fine-grained clastics from the Indus Fan and gravity flows from the Indian peninsula (Directorate General of Hydrocarbons).

2.6.3 Structural Features

Fault Systems

N-S transform faults with strike-slip kinematics (evidenced by seismic flower structures) control sediment distribution and slope stability. Active faulting in the Gulf of Kutch contributes to localized seismicity.

Basement Highs

The Vengurla Arch and Laxmi Ridge partition basins, creating uplift zones that influence foundation design for offshore infrastructure. The Laxmi Ridge, a continental sliver, separates the Laxmi Basin from the Saurashtra offshore basin.

Deepwater Morphology

Steep continental slopes (5° to 15°) transition into the Laxmi Basin, characterized by sediment gravity flows and overpressure zones (>12 MPa) in Eocene shales.

2.6.4 Sedimentary Basins

The offshore west coast hosts several sedimentary basins rich in hydrocarbons and other mineral resources.

These basins exhibit diverse stratigraphic sequences and depositional environments as summarized below in **Table 2.6.4-1.**

Basin Name	Location	Geological Features	Hydrocarbon Potential
Mumbai Offshore Basin	Extends offshore from Gujarat to Maharashtra.	Dominated by thick Mesozoic and Cenozoic sediments deposited in rift and post-rift settings.	Home to major oil and gas fields like Bombay High
Cambay Basin	Offshore extension of the onshore Cambay Basin near Gujarat.	Contains thick Tertiary and Quaternary sediments overlying Precambrian basement rocks.	Known for both onshore and offshore oil and gas reserves.
Kerala-Konkan Basin	Extends offshore from Karnataka to Kerala.	Characterized by deep-water sediments	Including turbidities and pelagic deposits.
Kutch Basin	Located near the Gulf of Kutch.	Highly faulted basin with significant structural complexity.	Limited exploration so far

Table 2.6.4-1: Summary of Sedimentary Basins in West Coast of India

2.7 Geotechnical Aspects

2.7.1 Soil Composition and Stratigraphy

The offshore soils along the west coast exhibit diverse compositions depending on their location and depositional environment:

Northern Regions (Gujarat):

- **Soil Types**: Predominantly silty clays and sandy silts, often weakly consolidated.
- **Engineering Properties:** Low shear strength, high compressibility, and susceptibility to liquefaction.

Central Regions (Maharashtra and Goa):

- **Soil Types:** Clayey silts and fine sands interbedded with organic-rich layers.
- **Engineering Properties:** Moderate to high plasticity indices; presence of gas hydrates complicates drilling operations.

Southern Regions (Karnataka and Kerala):

- **Soil Types:** Fine-grained sediments such as clay and silt, occasionally mixed with biogenic materials.

- **Engineering Properties:** High sensitivity to disturbance, low density, and susceptibility to liquefaction during seismic events.

2.7.2 Seismic Activity and Stability

While the west coast of India lies in a relatively stable tectonic zone compared to the eastern coast, localized seismicity occurs due to active faults within the continental shelf and subduction-related stresses at the Carlsberg Ridge. Seismic considerations are critical for designing offshore structures, especially in deeper waters where soft sediments amplify ground motion.

2.7.3 Challenges in Offshore Engineering

Several geotechnical challenges arise when developing infrastructure along the west coast:

- **Soft Sediments:** Require specialized foundation designs such as pin piles/skirts for shallow foundations. Spud cans penetrate deeper depths and risk of high uplift pressures.
- **Gas Hydrates:** Present in some areas, necessitating careful site investigation and mitigation strategies.
- **Monsoon Currents:** Strong seasonal currents cause scour around structures and affect seabed stability.
- **Corrosion:** High salinity levels accelerate corrosion of steel components, requiring corrosion-resistant materials and protective coatings.

2.7.4 Seafloor Morphology

The seafloor morphology varies significantly along the west coast:

- **Northern Regions (Gujarat):** Broad and shallow continental shelf influenced by fluvial systems like the Narmada and Tapti rivers.
- **Central Regions (Maharashtra and Goa):** Steep continental slope with numerous submarine canyons and channels.
- **Southern Regions (Karnataka and Kerala):** Narrower continental shelves and deeper waters, with fine-grained sediments dominating the abyssal plains.

2.7.5 Geotechnical Implications

Foundation Stability

- Horst-graben structures require site-specific designs to address differential subsidence.

- Lime mud layers (high $CaCO_3$ content) exhibit low shear strength (su=5–15 kPa), necessitating raft foundations for shallow structures.

Hydrocarbon Exploration

- Syn-rift carbonate reservoirs (e.g., Bassein Formation) require advanced coring techniques to assess fracture permeability.
- Overpressure zones in Eocene shales pose drilling risks, requiring managed-pressure drilling systems.

Seismic Hazards

Strike-slip fault reactivation can induce cyclic softening in clay-rich strata, affecting slope stability on the continental rise.

2.8 Case Studies

2.8.1 Case Study-1 (Dr Prashanth Talkad, Ms Minnu KP, Dr CR Parthasarathy, Mr Srinivas Nagappa & Mr Nishanth)

Introduction

Embarking on an offshore exploration journey off the east coast of India, we delve into one of the promising offshore site for the operator, where the challenges are equally significant, with water depths ranging from 20 to 3,200 meters.

This case study focuses on the crucial assessment of a jacket mudmat stability, employing the advanced Finite Element Analysis software, Plaxis 3D in shallow water.

Problem Statement

The offshore Jacket is temporarily supported by the skirted mudmat during installation prior to driving piles. During installation, the mud mat is subjected to combined vertical, horizontal and moment (VHM) loading, resulting from an eccentric gravity load and environmental loads.

Due to the presence of very soft Clay layer below the mud mat, the stability during installation would be influenced and hence needed to be evaluated. As the problem is three dimensional and cannot be analyzed with simplified approaches, finite element analysis is generally undertaken.

This case study deals with the mudmat-soil interaction modelling using Plaxi-3D for estimating the mudmat settlements and tilt.

Project Details

A living quarters and utility platform (LQUP) was planned to be installed off the East Coast of India at a water depth of 30m. The skirted mudmat for this location comprised four (4) pile slots viz., A2, A5, B2 and B5 as shown in **Figure 2.8.1-1.** A skirt of 1.5m deep along the periphery of mudmat was designed.

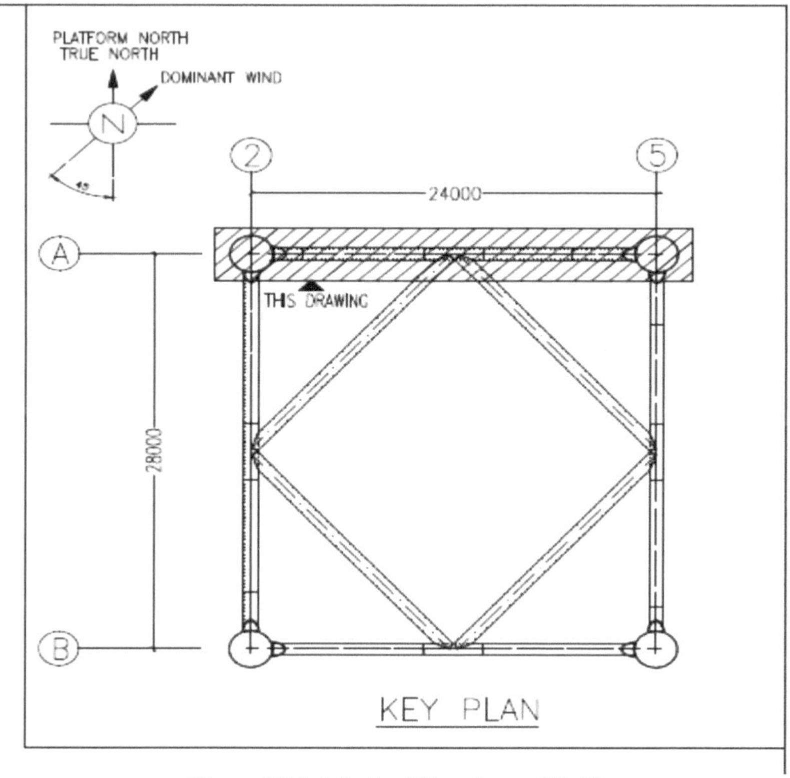

Figure 2.8.2-1: Jacket Elevation at Mudline

Soil Conditions

A site-specific geotechnical investigation was carried out with four deep boreholes with final depths ranging from 71m to 160m. Geotechnical investigation report indicates that the mudline stratigraphy comprises Very Soft Clay. Both in-situ and laboratory shear strength data indicate that the undisturbed shear strength of this clay layer is less than 5kPa up to a depth of approximately 2m below the mudline. The bottom of the mud mat's peripheral skirts is designed to rest at a depth of 1.5m below the mudline.

Given the presence of this highly deformable clay layer beneath the mud mat, ensuring stability during installation was essential. Throughout the installation process, the mudmat will experience a combination of vertical, horizontal, and moment (VHM) loading, resulting from eccentric gravity loads and environmental influences.

Due to the complex three-dimensional nature of the situation, simplified or two-dimensional analyses are inadequate. Consequently, Finite Element Analysis was performed using PLAXIS 3D software to provide a precise evaluation of the scenario.

The modulus of elasticity of Clay layers varies between 100*Su to 500*Su as per the available literature (Bowles, 1996). For this project, a value of 150*Su was considered. A very conservative value of 150*Su aligns with the minimum value of 5*q_c (Bowles, 1996). S_u and q_c are undrained shear strength and CPT cone tip resistance of Clay, respectively. Based on this, the soil design profile presented in **Table 2.8.1-1** was developed for modelling in Plaxis 3D.

Layer	Depth (m)	Unit Weight (kN/m3)	Undrained Shear Strength, Su (kPa)		Modulus of Elasticity, Es (kPa)		Poisson's ratio (μ)
			(Top)	(Bottom)	(Top)	(Bottom)	
1	0 - 1	14.0	1	2	150	300	0.49
2	1 - 2	15.0	2	3	300	450	0.49
3	2 - 3	15.0	3	5	450	750	0.49
4	3 - 4	16.0	5	6	750	900	0.49
5	4 - 6	16.0	6	9	900	1350	0.49
6	6 - 7	16.5	9	11	1350	1650	0.49
7	7 - 14	16.5	11	14	1650	2100	0.49
8	14 - 19	17.5	14		2100		0.49
9	19 - 25	17.5	20		3000		0.49
10	25 - 30	18.0	27		4050		0.49
11	30 - 52	19.0	59		8850		0.49
*Note: 0.49 is the default value of Poisson's Ratio for Undrained C model in Plaxis 3D							

Table 2.8.1-1: Soil design parameters

Load Combinations

In addition to the top side loads on piles, the environmental loads in different directions lead to several load combinations for analysis.

Case	Load Combination	Node Location A2					
		Fx (kN)	Fy (kN)	Fz (kN)	Mx (kN-m)	My (kN-m)	Mz (kN-m)
Base Jacket	1043	-950.29	293.17	-5515.91	-947.96	-2995.63	204.92
		Node Location A5					
		Fx (kN)	Fy (kN)	Fz (kN)	Mx (kN-m)	My (kN-m)	Mz (kN-m)
		-394.82	242.64	-3137.93	-543.97	-1159.4	133.71
		Node Location B2					
		Fx (kN)	Fy (kN)	Fz (kN)	Mx (kN-m)	My (kN-m)	Mz (kN-m)
		-989.1	-292.35	-5694.32	983.8	-3124.55	-144.92
		Node Location B5					
		Fx (kN)	Fy (kN)	Fz (kN)	Mx (kN-m)	My (kN-m)	Mz (kN-m)
		-409.47	-240.28	-3246.6	593.63	-1186.85	-164.17

Table 2.8.1-2: Critical load combinations for load combination Base Jacket

Case	Load Combination	Node Location A2					
		Fx (kN)	Fy (kN)	Fz (kN)	Mx (kN-m)	My (kN-m)	Mz (kN-m)
Base Jacket	1017	-748.32	-144.69	-5544.58	341.47	-2465.22	379.76
		Node Location A5					
		Fx (kN)	Fy (kN)	Fz (kN)	Mx (kN-m)	My (kN-m)	Mz (kN-m)
		-253.25	-29.863	-3588.49	209.7	-675.87	29.43
		Node Location B2					
		Fx (kN)	Fy (kN)	Fz (kN)	Mx (kN-m)	My (kN-m)	Mz (kN-m)
		-928.63	-691.69	-6778.7	2357.35	-3051.27	108.48
		Node Location B5					
		Fx (kN)	Fy (kN)	Fz (kN)	Mx (kN-m)	My (kN-m)	Mz (kN-m)
		-375.14	-507.99	-4442.38	1579.34	-981.53	-462.814

Table 2.8.1-3: Critical load combinations for load combination Base Jacket

Case	Load Combination	Node Location A2					
		Fx (kN)	Fy (kN)	Fz (kN)	Mx (kN-m)	My (kN-m)	Mz (kN-m)
P1@ all 4 Legs + P2@B2 + P2@A2	2417	-821.07	-55.34	-9317.23	-464.19	-3261.73	370.48
		Node Location A5					
		Fx (kN)	Fy (kN)	Fz (kN)	Mx (kN-m)	My (kN-m)	Mz (kN-m)
		-198.98	-21.04	-4546.62	55.47	-529.58	22.16
		Node Location B2					
		Fx (kN)	Fy (kN)	Fz (kN)	Mx (kN-m)	My (kN-m)	Mz (kN-m)
		-993.68	-790.24	-10577.11	3193.91	-3837.08	129.27
		Node Location B5					
		Fx (kN)	Fy (kN)	Fz (kN)	Mx (kN-m)	My (kN-m)	Mz (kN-m)
		-317.95	524.77	-5463.4	1774.21	-804.01	-456.46

Table 2.8.1-4: Critical load combinations for load combination Base Jacket

The structural software will identify the critical load combinations. In this project, three critical load combinations were identified by the structural team and the same are considered for analysis. These are summarized in **Tables 2.8.1-2, 2.8.1-3 and 2.8.1-4** as shown above.

Problem Geometry

The lateral dimensions of the 3D model are 120 x 130m, while the model thickness is 52m with the soil stratigraphy described in **Table 2.8.1-1**. These dimensions were selected such that the boundary effects are negligible. Based on the mud mat geometry provided, the skirts, mud mat & yoke plate were created in PLAXIS 3D. The depth of the skirts was 1.5m. The jacket mud mat along with skirts were modelled as a rigid plain plate with steel properties.

FEM Model

The soil beneath the mud mat is treated as an isotropic Elasto-Perfectly plastic continuum governed by the Mohr-Coulomb yield criterion. Due to a lack of strength test data close to the mud mat (within 10m), a simpler elasto-plastic continuum model is used instead of a more advanced soil model. The clay layers are assumed to behave as undrained, with cohesion equal to the undrained shear strength (S_u). Elastic behavior is characterized by a Poisson's ratio of 0.49 and a constant ratio of Young's modulus to undrained shear strength (E/S_u) of 150.

The interface strength between the mud mat and clay layers is modeled using an interface factor 'a', where maximum shear stress equals the undrained shear strength. Different levels of interface roughness are considered: "rough" (a value of 1), "smooth" (a value of 0), and an intermediate value of 0.7, typical for a steel/soft clay interface.

The jacket mud mat with skirts is modeled as a rigid plain plate, and the seabed is assumed to be perfectly flat.

The mesh for analysis consists of 10-noded tetrahedral elements. Initially set to a "Medium" global coarseness, the local refinement factor decreased to 0.5 in the upper clay layer where failure mechanisms are expected to develop.

Phased Calculations

Initial Phase: The Initial Phase represents soil layers with no influence of mudmat with skirts.

Phase 1: In this phase, Mudmat with skirts and its interface are introduced.

Phase 2: In this phase all the loads are activated.

Sample Result and Summary

Figure 2.8.1-2 is a sample output of total displacement under load combination of 1017, while **Figure 2.8.1-3** is the geometry of mud mat used for analysis.

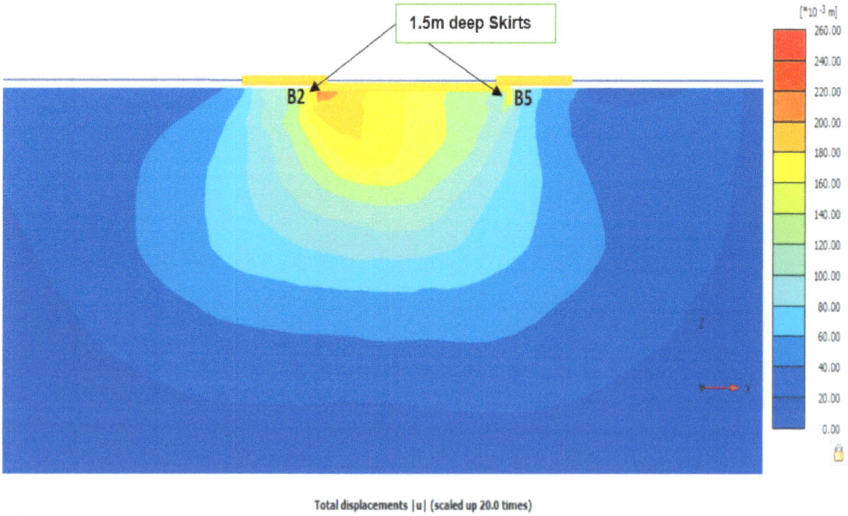

Figure 2.8.1-2: Total displacements under Load Combination 1017

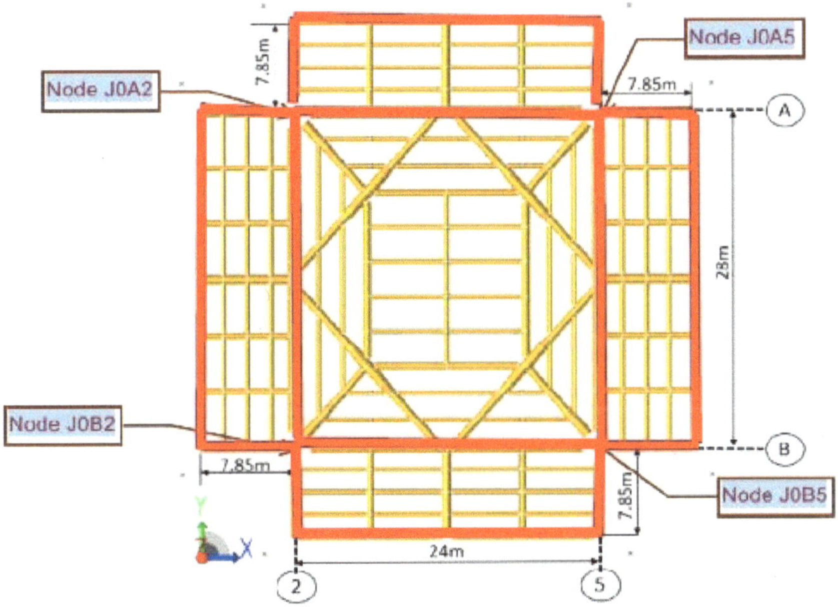

Figure 2.8.1-3 : Geometry of Mudmat used for analysis

For load Combination 1017, Location B2 experiences more vertical settlement compared to other leg locations **Table 2.8.1-5.** This was compared by performing analytical estimations of the mudmat settlement. The tilt from Plaxis analysis for this load combination of < 0.3°.

All Displacements in mm											
A2 Location			A5 Location			B2 Location			B5 Location		
Ux	Uy	Uz	Ux	Uy	Uz	Ux	Uy	Uz	Ux	Uy	Uz
40	25	176	40	25	71	42	25	250	42	27	128

Table 2.8.1-5: Axial displacements in mm under Load Combination 1017

Similarly, under load combination 1043, Location A2 & B2 experiences near uniform vertical settlement compared to other leg locations as shown in the **Table 2.8.1-6,** below. The tilt estimated was less than 0.3°.

All Displacements in mm											
A2 Location			A5 Location			B2 Location			B5 Location		
Ux	Uy	Uz	Ux	Uy	Uz	Ux	Uy	Uz	Ux	Uy	Uz
47	0.6	188	47	0.1	70	48	0.5	200	48	0.3	76

Table 2.8.1-6 : Axial displacements in mm under Load Combination 1043

And under load combination 2417, Location A2 & B2 experiences more vertical settlement compared to other leg locations as shown in **Table 2.8.1-7.** The tilt estimated from plaxis analysis was around 0.52°.

All Displacements in mm											
A2 Location			A5 Location			B2 Location			B5 Location		
Ux	Uy	Uz	Ux	Uy	Uz	Ux	Uy	Uz	Ux	Uy	Uz
45	31	338	45	30	80	45	31	425	43	30	145

Table 2.8.1-7 : Axial displacements in mm under Load Combination 2417

The above analyses were also performed with the consideration of reduced soil modulus values. The same three load combinations 1017, 1043 and 2417 were applied to two cases where the soil modulus was estimated as 100*Su and 50*Su.

The tilt under load combination 1017 under the above-mentioned cases increased to 0.41° and 0.81° respectively. Similarly, under the load combination 1043, the tilt increased to 0.31° and 0.62°. And, under the load combination 2417, the tilt increased to 0.77° and 1.53° all moving slightly away from the A2-B2 section.

Conclusions

The mudmat-soil interaction was modelled successfully in Plaxis-3D and the mudmat settlement and the tilt were determined for the critical load combinations analyzed. The maximum vertical settlement determined was 425mm and the corresponding tilt was 0.52°.

The sensitivity analyses carried out for soil modulus values of 100*Su and 50*Su resulted in a maximum tilt of 0.77° and 1.53°.

Finally, when the jacket was installed successfully. The observed settlement and tilt were similar to the results of the ground model corresponding to 150*Su.

2.8.2 Case Study -2 (Dr Prashanth Talkad, Ms. Ashitha, Minnu KP and Dr CR Parthasarathy)

Introduction

A Jack up rig was to move into a few platform locations in offshore Yanam, East coast of India located within Krishna Sub Basin, Andhra Pradesh, East Coast of India. This field has eight (8) unmanned well platforms, an onshore process plant, just over a km from the coast and an SBM with PLEM for the export of Oil. A Jack-up rig suitability was to be assessed for moving near one of the existing platforms.

Problem Statement

Jack up rig suitability study is necessary prior to moving into the location. The study includes assessment of leg penetration, punch-through potential, rig stability analysis for V-H-M loading, effect of scouring and spud can interaction assessment with the existing Jacket foundation. This case study discusses the methodologies for bearing; sliding and overturning stability checks of a Jack up rig planned to move into an offshore site next to an existing Jacket platform in East Coast of India. The scour potential and the load required for leg extraction were also analyzed. The spud can interaction with the existing nearby platform was modelled in Plaxis-3D and the analyses results are discussed.

Project Details

A Jack up rig was planned to move into an offshore site next to an existing Jacket platform.

The Jack up rig details are shown in **Table 2.8.2-1.** The existing platform was a Tripod Jacket having piles battered in both the rows. Site-specific Geophysical and Geotechnical investigations have been conducted.

Description	Unit	Value
Total length of leg	m	145.39
Tip of spudcan to widest section	m	3.2
Equivalent diameter	m	14.02
Equivalent footing area of each spudcan	m^2	148.27
Preload per leg	MN	53.38
Lightship load per leg	MN	31.7
Diameter of conical tip	m^2	0.82

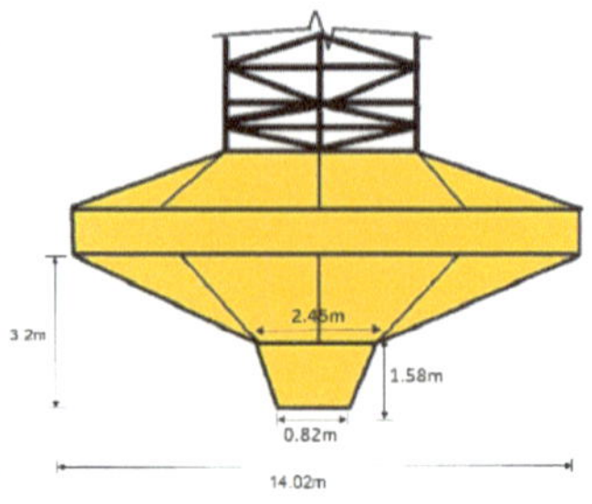

Table 2.8.2-1: Jack-up Rig details

Review of Geophysical Data

General bathymetry at current location presented an almost smooth and flat seabed with a regional gradient towards south. Maximum water depth within survey area was recorded as 11.2m. The seabed is gently sloping towards Northeast to Southwest having a gradient of 1.720 (0.08°).

Side scan sonar data suggested the seabed to have varying reflective frequencies which can be broadly categorized as Type-1 (soft to medium silty clays) and Type-2 (Stiff Clays with admixture of Sand) sediments. There were three existing pipelines in the vicinity of this platform location. Apart from the above, no other features of significance/items of debris were recorded within the survey corridor, which could be hazardous to existing pipeline route and Jack up emplacement.

Review of Geotechnical Data

Review of geotechnical investigation report indicated that the field investigation consists of drilling and sampling of exploratory borehole of 100.9m depth. The soil stratigraphy at the proposed jack-up rig location predominantly consists of CLAY with intermittent Sand layers up to a borehole depth of 82m followed by Sand up to100.9 m depth.

Notable, a 2.5m thick SAND layer is present at 16m, with in the cohesive deposit. The consistency of clay layers is generally Very Soft to Stiff at initial depths and Very Stiff to Hard at deeper stratum.

The water depth at the proposed jack-up rig location is 11.2m. **Table 2.8.2-2** summarizes the design ground model adopted.

Sl. No.	Vertical depth (m)		Soil Type	Undrained Shear Strength (kPa)		Effective Unit Weight (kN/m3)	Angle of Internal Friction (ϕ), deg	
	From	To		LB	UB		LB	UB
1	0.0	5.5	Clay	6	8	6.0	-	-
2	5.5	16.0	Clay	15	25	6.0	-	-
3	16.0	18.5	Sand	-	-	6.5	25	25
4	18.5	30	Clay	36	44	6.5	-	-
5	30.0	41.0	Clay	50	75	6.5	-	-
6	41.0	68.0	Clay	75	125	7.0	-	-
7	68.0	73.0	Sand	-	-	9.0	30	30
8	73.0	82.0	Clay	200	200	9.5	-	-

Table 2.8.2-2: Soil design parameters

Review of Metocean Data

Based on the metocean data and rig details, the environmental loadings such as wind force, current force and drift force have been computed by the Naval Architects for this Jack up rig and the same has been shown in **Table 2.8.2-3.**

Description	1 Year Returning Period	10 Year Returning Period	50 Year Returning Period
Total wind force on the jack-up rig (Longitudinal Direction) (MN)	1.08	5.41	10.52
Total wind force on the jack up rig (Transverse Direction) (MN)	1.08	5.42	10.55
Total current force on the jack-up rig (MN)	0.17	0.30	0.41
Total drift force on the jack-up rig (MN)	4.91	6.64	7.28
Total force (Longitudinal Direction) (MN)	**6.17**	**12.36**	**18.22**
Total force (Transverse Direction) (MN)	**6.17**	**12.37**	**18.24**

Description	1 Year Returning Period	10 Year Returning Period	50 Year Returning Period
Moment due to wind force (Longitudinal Direction) (MN-m)	56.15	280.57	545.80
Moment due to wind force (Transverse Direction) (MN-m)	56.74	283.51	551.52
Moment due to current force (MN-m)	0.97	1.70	2.32
Moment due to drift force (MN-m)	55.02	74.42	81.54
Total overturning moment (Longitudinal Direction) (MN-m)	**112.15**	**356.70**	**629.67**
Total overturning moment (Transverse Direction) (MN-m)	**112.74**	**359.64**	**635.38**

Table 2.8.2-3: Summary of environmental loads

Leg Penetration Analysis (SNAME 2008, ISO 19905-2023)

The spud cans of jack-up rigs are typically treated as shallow foundations. This analysis is based on **ISO 19905-1, SNAME** guidelines and general bearing capacity theory appropriate to the soil types derived for the proposed rig move location. The results of the analysis are presented in the **Table 2.8.2-4 and Figure 2.8.2-1.**

Location	Tip penetration(m)	Remarks
Jacket platform Location 1	**28.0m to 33.0m** (Preloading of 53.38 MN)	"Punch Through" potential (of SAND layer at 16 to 18.5m and underlying weak CLAY) of the Jack up rig spud can was **INDICATED** (medium to High probability) at this **Jacket platform** location during full preloading (53MN). Rapid leg penetration was predicted between 19 to 33m. Hence, it was recommended to adopt necessary precautions while lowering the jack up rig legs. The preloading to be carried out on individual legs.

Table 2.8.2-4: Leg penetration analysis results

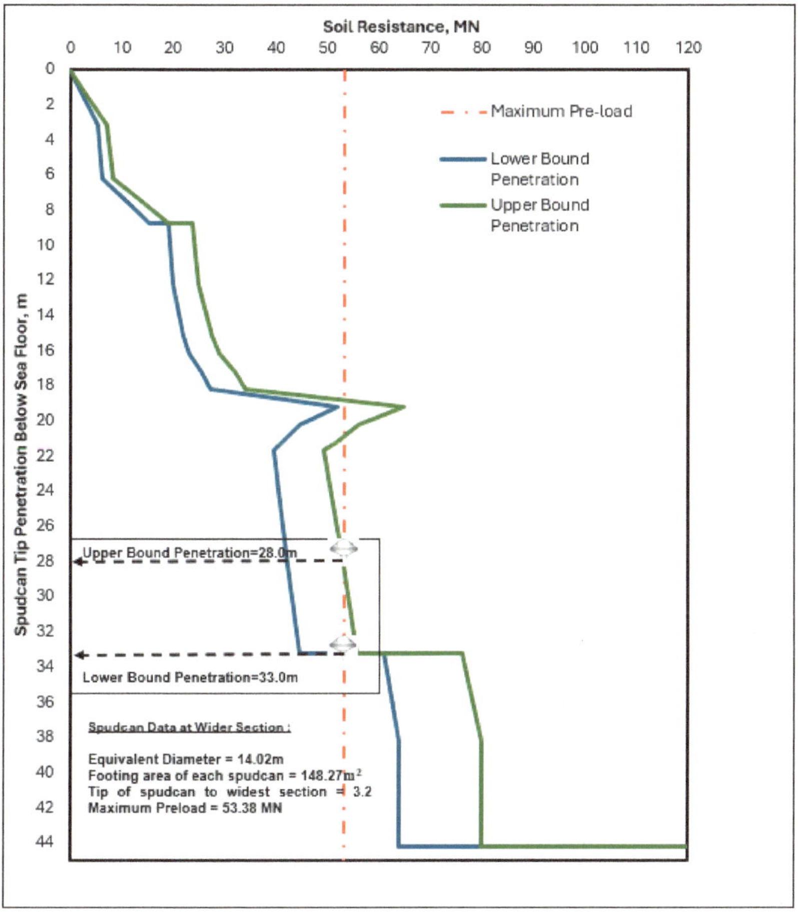

Figure 2.8.2-1 : Spud can penetration results

Foundation Bearing Check

Foundation Bearing capacity (BC) of the founding soil is checked for the combined effect of VHM loads acting on the spud cans. These loads are in operation for 1 year return period, Preload for 10 year return period, and maximum storm holding load for 50 year return period. Conventional shallow Bearing Capacity equations with relevant reduction factors as per **API 2GEO (R2021)** are considered in the analysis.

Foundation Bearing check is performed for the rig at penetration layer (33 m for 1 & 10 year return period at 53MN Load). The factored V-H envelope (foundation bearing safety factor) was developed based on procedures outlined in Clarom (Ref 13) and SNAME (2008).

The bearing checks resulted in a FOS of 1.30 and 1.2 for 1 year and 10 years return periods respectively. It was found that critical loads are within the bearing envelope, hence any foundation bearing issues were not envisaged for this rig at this location for 1 year & 10-year return period. The V-H envelop plots are presented in **Figure 2.8.2-2 and 2.8.2-3.**

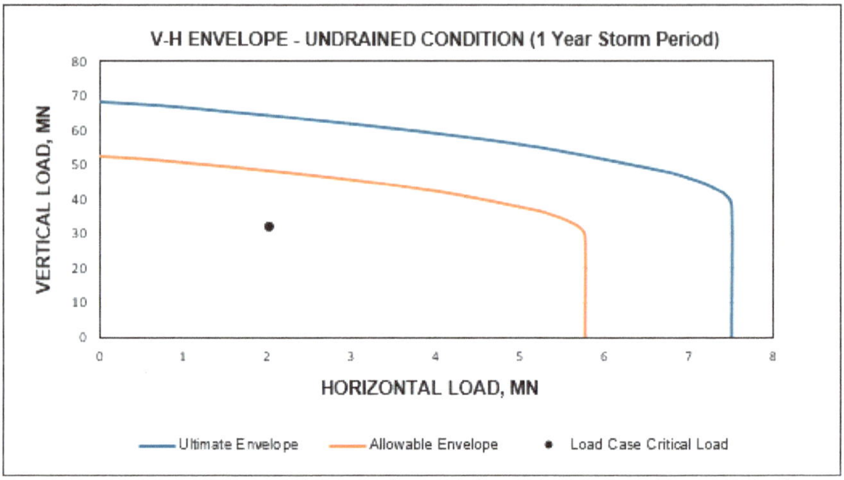

Figure2.8.2-2: V-H Envelope-Undrained Condition (1 year Storm Period)

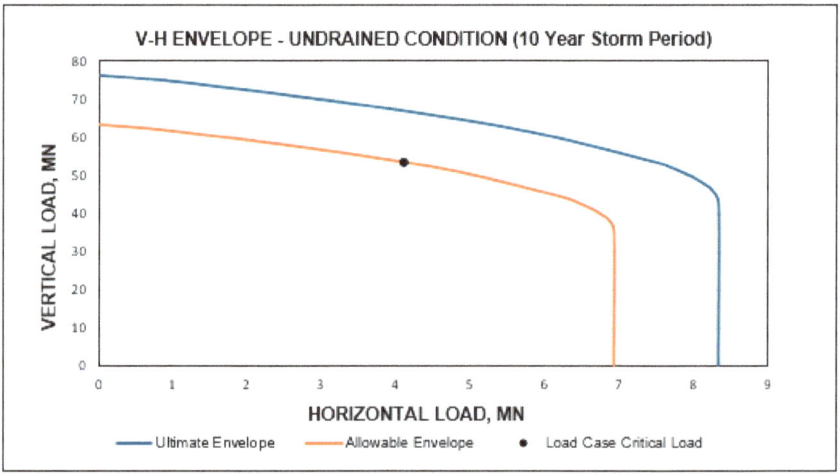

Figure 2.8.2-3: V-H Envelope-Undrained Condition (10 year Storm Period)

For the rig to satisfy the bearing check for 50year return period, the rig had to be prepared for full storm conditions by loading to 60 MN per leg. For the 50 year storm period, the maximum VHM conditions were also well within the ultimate bearing envelope.

This is in line with the practice of the industry to check the extreme storm conditions to ultimate resistance of the soil. The V-H envelope plots are presented in **Figure 2.8.2-4**.

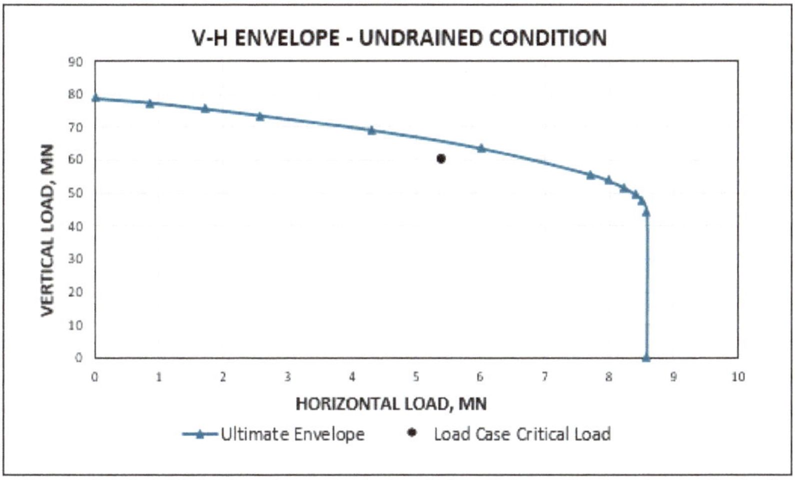

Figure 2.8.2-4: V-H Envelope-Undrained Condition (50 Year Storm Period)

Horizontal Sliding Stability

The sliding stability criterion of jack-up rig penetrated through clay layer was evaluated based on the formulae provided in Clarom. Based on the design soil profile as per **Table 2.8.2-2**, the sliding stability was evaluated, and the safety factor was more than 1.3 for this rig at this **location** for 1-year, 10-year and 50-year return period of environmental loads. Hence the **jack up rig was safe against sliding at this location.**

Stability Check for Overturning

During operation, the deck and legs of the Jack up rig are subjected to horizontal loads (due to wave, wind and drift) which cause an overturning effect (moment) on the jack-up rig. The stability checks on overturning (FOS) are intended to ensure adequate safety against uplift of the Jack up leg(s). The weight of the jack-up rig on the legs shall be higher than the uplift force induced due to environmental forces, to prevent overturning.

Gravity loading (i.e. lightship load) at operating conditions is distributed equally among the three legs of Jack-up rig.

The environmental loads and corresponding overturning moments on the jack-up rig were considered based on the metocean data and are summarized in **Table 2.8.2-5.**

The forces due to wind, wave & drift are considered to act in the same direction (i.e., for the most unfavorable case). Corresponding moments arising due to each of these environmental forces are then added to get the total overturning moment. To arrive at factored overturning moment, DAF of 1.5 (conservative estimate) is multiplied to the overturning moment.

The restoring moment (considering the jack-up rig to be rigid) is a product of reaction under the leg, multiplied by the distance between the leg & the axis of the other two legs with appropriate reduction due to the following, P_{Δ} effect and Buckling risk of legs. The FOS against overturning is the ratio of restoring to the factored overturning moment. This value shall be more than 1.25 for the rig to be stable (against overturning).

Overturning stability was checked for the current location and found to be safe for 1-year, 10-year and 50-year return period. **Table 2.8.2-5** summarizes the details of the overturning stability check.

Description	Units	Storm Periods		
		1 Year Return Period	**10 Year Return Period**	**50 Year Return Period**
Factored overturning moment	MN-m/ leg	37.58	119.88	211.79
Net restoring moment	MN-m/ leg	1152.91	1152.91	1152.91
F.O.S	-	**30.68**	**9.62**	**5.44**
Note- The maximum preload considered is 53 MN.				

Table 2.8.2.5: Overturning stability check results

Preload Assessment Criteria

The preload assessment criterion recommended per leg was computed as per Clarom and shown in **Table 2.8.2-6.**

Description	Units	1 Year Return Period	10 Year Return Period	50 Year Return Period
Permanent load per leg (lightship load per leg at zero air gap is considered as permanent load per leg), P	MN	31.70	31.70	31.70

Description	Units	1 Year Return Period	10 Year Return Period	50 Year Return Period
Moment per leg due to horizontally acting environmental loads, M	MN-m	37.58	119.88	211.79
Lever Arm = Distance between bow leg & axis of other 2 legs	M	43.28	43.28	43.28
Vertical environmental load per leg, Ev (1)	MN	0.86	2.76	4.89
Variable load per leg, L (2)	MN	11.22	11.22	11.22
Gravity load per leg, W = P + L	MN	42.92	42.92	42.92
(ϒW * W) + (ϒE * Ev)	MN	44.124	46.78	49.76
Preload per leg, QP	MN	53.00	53.00	53.00
Preload assessment criterion satisfied		YES	YES	YES

Table 2.8.2-6: Preload assessment criterion (Clarom)

Leg Extraction Analysis:

The procedure to pull out Jack-up spud cans depends on the extent of its penetration below the mudline. Jacking systems generally prove sufficient to pull out spud cans that have penetrated less than 10.0m (low penetrated spud cans). If significant resistance is encountered during pull-out, the hull can be further lowered below the normal floating draft position to facilitate pull out. The pull-out force was estimated using the following expression as per Clarom (See page 70). If the penetration exceeds 10.0m (deep penetrated spud cans), it is quite likely that the pull-out force required will exceed the capacity of jacking system. In such cases, pull-out must be assisted with jetting on the underside and upper side of the spud can. Jetting on the underside is necessary to break the suction force (and to shear the soil/spud can adhesion in clayey soils). Jetting on the upper side is necessary to reduce the shear strength of the accumulated soil (i.e. the soil deposited after lowering the spud can).

Total pull-out force/ leg (Q_P) required was 41.05MN which was greater than the available Jacking system capacity of 25.33MN/leg (as per the rig operating manual). Hence, jetting in the spud cans were recommended for the Jack-up rig at this platform location to release base suction.

Scour Evaluation and Mitigation Plan

Typically, the scoring phenomena is influenced by velocity of water current at seabed, water depth and soil particle size. Scouring is said to occur if the velocity of water particles at seabed is greater than the critical soil particle velocity. Velocity of water particles at seabed is the summation of orbital velocity (V_{orb} depends on significant wave height, water depth and time period) and water current velocity at seabed (V_c measured at site) as per Clarom.

$$V_{total} = V_{orbital} + V_{current\ soil} \qquad \textit{Equation 2.8.2-1}$$

From the metocean data and geotechnical soil borelogs, the critical velocity U_c was moderately lower than the velocity of water particles. The individual spudcan legs of Jackup rig at this **platform** location has a significant penetration in the order of 28m (minimum) and **hence, chances of minimal scouring were envisaged at this platform location. This was also in-line with the observed scour data (furnished by client).**

Evaluation of Challenges for Rig Positioning and Mitigation Plan

Based on the review of available platform and pipeline drawings, the orientation of the existing platform is towards North-East direction and the network of gas pipelines connecting the platform is towards the South-West direction. **The conductor at this platform has eleven slots oriented towards the North-East direction.**

The platform is supported by two rows of piles and the first row of piles towards the conductor are vertical and the other side are battered. The inference from the drawings also suggests that the **exposed pipelines at this location do not pose any significant risk for positioning the Jackup rig.** Also, the geophysical data suggest the seabed surface at this platform location is gently sloping at angle of 0.08° towards South-West from North-East direction. Hence the chances of rig tilt were also not envisaged at this location due to deeper penetration of the Jack up legs.

Jacket Spud can Interaction

Based on the review of as-built platform drawings, the four conductor slots of the Platform are oriented along North- East direction. The jacket piles of the platform are double battered, and the Jack-up rig was planned to be positioned towards the conductors. As a result, a clearance of 17m was envisaged between the spud can and the platform pile at the depth of 28m below the mudline which was more than one diameter of the jack-up spud can. The cantilever portion (Skid unit cantilever) of Jack-up rig could reach the farthest conductor slots comfortably for drilling.

Penetration Effect of Spud cans on the Nearby Jacket

Figure 2.8.2-5 shows the penetration and proximity of spud can with the Jacket platform pile.

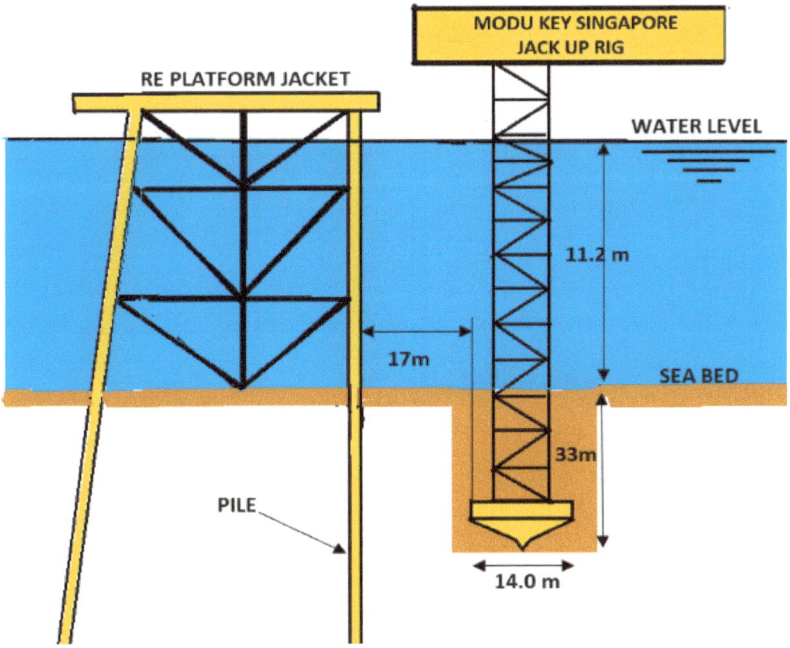

Figure 2.8.2-5.: Proximity of spud can with the Jacket

Typically, in soft to stiff clays the penetration of spud can results in a highly disturbed influence zone of 2 times diameter of spud can as indicated in **Figure 2.8.2-6.**

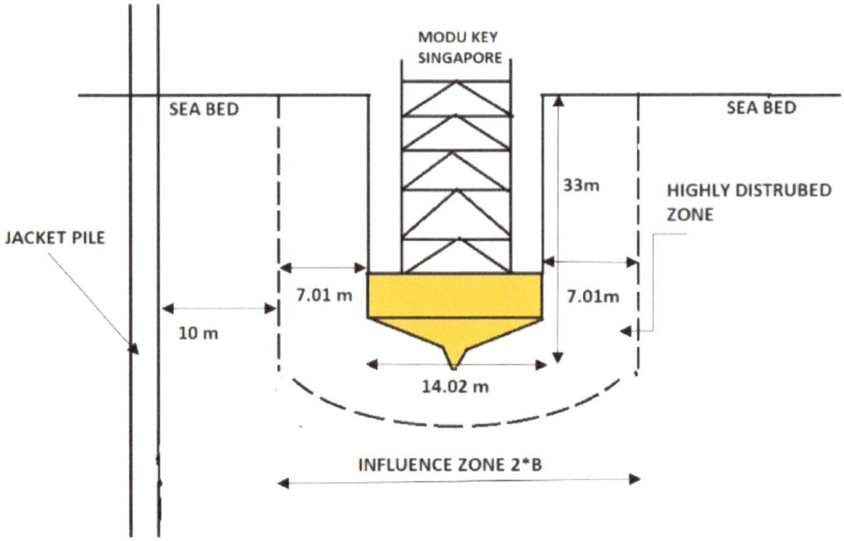

Figure 2.8.2-6 Spudcan influence zone

It is evident from **Figure 2.8.2-6** that the Jacket platform pile is **10.0m** away from the highly influence zone of spud can penetration. Hence, the penetration of the spud can by itself will not induce additional stress on the existing platform Jacket piles. This finding was in line with Section 6.4.6 of **SNAME 2008** that the influence of spud can penetration beyond one diameter of spud can has no bearing on the existing platform jacket pile. However, during the leg extraction of the spud can, a reduction of undrained shear strength of clay layers surrounding the existing jacket piles up to a depth of 33m was expected. This reduction of shear strength along the pile (up to 33m from mudline) was captured in the form of P-Y springs as per **API RP2A GEO (R2021)**. The average undrained shear strength at platform location up to a depth of 33m was approximately 26KPa. From the furnished geotechnical data, laboratory test measurements at nearby platform locations suggest a sensitivity of Clay to be about 1.5 to 2.3. In computing the reduced p-y springs, a sensitivity of 2.5 which was also a conservative estimate was adopted in the interaction study. This highly conservative remolded shear strength was adopted to capture the degradation due to jetting process during leg extraction. The same was incorporated in the SACS analysis to evaluate the stability and integrity of the existing jacket piles. The SACS analysis was carried out for **undisturbed soils + operating storm (1year Return period) / Extreme storm (100year return period)** for in-place analysis of the platform during spud can penetration. For the condition of leg extraction of Jack up rig, **disturbed soils + operating storm (1year Return period) / Extreme storm (100year return period),** was considered.

The values of SACS analysis for both undisturbed and disturbed soil condition are presented in **Table 2.8.2-7** and **Table 2.8.2-8** respectively. The F.O.S meets the requirement for the degraded soil springs up to 27m and the displacements are also within the prescribed limits.

Pile (Grid)	Undisturbed Soil							
	Operating storm				Extreme storm			
	FOS	Status	Max Pile UC		FOS	Status	Max Pile UC	
			Above Mud Line	Below Mud Line			Above Mud Line	Below Mud Line
A1	4.25	>2.0, OK	0.456	0.389	2.30	>1.5,OK	0.657	0.533
A2	4.07	>2.0, OK	0.467	0.395	2.25	>1.5,OK	0.669	0.575
B1	3.88	>2.0, OK	0.519	0.457	2.04	>1.5,OK	0.839	0.751
B2	3.85	>2.0, OK	0.519	0.457	2.02	>1.5,OK	0.846	0.763

Table 2.8.2-7: SACS results for undisturbed soil condition at platform location (Undisturbed soil + Operating storm (1year RP) / Extreme storm (100year RP)

Pile (Grid)	Disturbed Soil (Sensitivity 2.5)							
	Operating storm				Extreme storm			
	FOS	Status	Max Pile UC		FOS	Status	Max Pile UC	
			Above Mud Line	Below Mud Line			Above Mud Line	Below Mud Line
A1	4.21	>2.0, OK	0.431	0.363	2.29	>1.5,OK	0.615	0.533
A2	4.04	>2.0, OK	0.441	0.368	2.24	>1.5,OK	0.626	0.545
B1	3.87	>2.0, OK	0.530	0.471	2.04	>1.5,OK	0.860	0.775
B2	3.84	>2.0, OK	0.530	0.471	2.02	>1.5,OK	0.867	0.787

Table 2.8.2-8: SACS results for disturbed soil condition at platform location (Disturbed soil up to 33m with degraded P-Y + Operating storm (1year RP) / Extreme storm (100year RP)

The summary of the SACS results clearly indicates that the leg extraction of the spud cans at the platform location had no impact on the degradation of the existing factor of safety for the in-place condition of the platform.

A finite element analysis of Jacket pile and spud can penetration was carried out using commercially available PLAXIS 3D software.

The geometry of the spud can, the Jacket platform pile and soil stratigraphy were considered as appropriate for this location. The soil layers were modelled using Mohr-Coulomb approach which is a typical linearly-elastic, perfectly plastic soil model. The clay layers were modelled as undrained condition while the sand layers were modelled as drained condition.

The Jacket pile was modelled as plate elements with varying wall thickness as per pile make-up drawing. The spud can was modelled as a weightless elastic (non-porous) rigid body with equivalent width of 14.02m with full base contact. The finite element mesh was generated using 15 node triangular elements. The initial geostatic conditions were calculated first followed by the pile placement of jacket and the spud can penetration. **Figure 2.8.2-7** shows the spud can penetration at 33m below the soil bed and the soil disturbance caused during the rig placement.

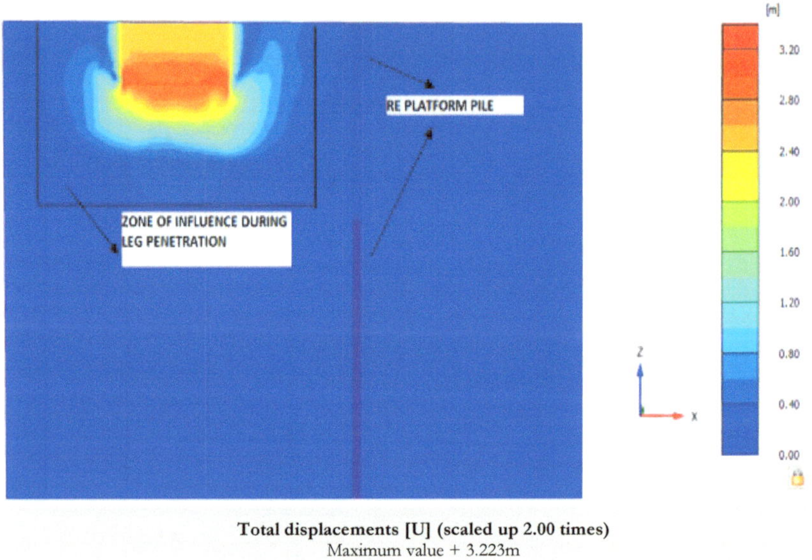

Figure 2.8.2-7: Spudcan and Jacket pile interaction in PLAXIS 3D

This clearly demonstrated that the highly disturbed zone due to leg penetration was away from the Platform Jacket pile. Hence, the stability of the existing jacket pile was not influenced due to leg penetration. This validated the FOS analyzed by SACS modelling.

Conclusions and Recommendations

- "Punch Through" potential (of SAND layer at 16.0m-18.5m and underlying weak CLAY) of the Jack up rig spud can was **INDICATED** at this platform location during full preloading (53MN). Rapid leg penetration was predicted to be between 19-33m. To mitigate the punch through hazards, it was recommended to conduct single leg preloading (This will also reduce the overall topsides load and therefore lessen the individual leg loads and leaves the other legs free to jack if required) while keeping the hull close to water and extreme caution shall be exercised while preloading.

- The FOS against horizontal sliding stability for the Jackup rig at this platform location was more than 1.3. Hence, the sliding failure was not envisaged for this rig at this location.
- The bearing checks resulted in a FOS of 1.30 and 1.2 for 1 year and 10 years respectively. It was found that critical loads were within the bearing envelope, hence any foundation bearing issues were not envisaged for this rig at this platform location for 1 year & 10-year return period.
- There was negligible effect of scouring at this platform location for this Jack-up rig. This was also in-line with the observed scour data (furnished by client).
- The total pull-out force/ leg (Q_p) required was more than the available jacking system capacity per leg. Hence, jetting was recommended to release base suction.
- The SACS analyses indicated that the leg extraction of the spud cans at the platform location would have no impact on the degradation of the existing factor of safety for the in-place condition of the platform.

Finally, the rig was moved to the location, and the observed leg penetration was about 88-91 feet, which was in close agreement to the predicted penetration.

2.8.3 Case Study-3 (Mr Srinivas Nagappa, Dr Gangisetti Ramesh, Ms Anusha Bekal, Dr CR Parthasarathy and Dr. Brent Robinson)

Problem statement:

Offshore pile driving may experience stoppages due to hammer malfunction and the resumption of driving after the repair or using backup hammer may not move the pile due to high soil setup development during the delay particularly if the stoppage occurs at deeper penetrations.

Pile adequacy check is done by estimating the available capacity for the penetration achieved. Hindcast analysis is performed to estimate the available pile capacity by modelling hammer-pile-soil interaction in

GRLWEAP software and matching the computed blow counts with the observed field values.

However, in cases where pile doesn't move during the restart, hindcast analysis under predicts the available capacity particularly the end bearing.

Pile instrumentation during the restart will aid in capturing the force-velocity signatures so that a CAPWAP analysis can be performed to arrive at the available skin and toe resistances.

The estimated toe resistance of the end of drive condition can be added to skin resistance during restrike test to arrive at a reasonable available total capacity. This case study examines a specific instance where offshore pile driving was halted due to hammer malfunction, and the pile adequacy assessments were made.

Project Details:

A four-legged Jacket platform was planned to be installed in west coast of India. The platform piles consisted of 2.134m OD open ended steel pipe piles which have variable wall thickness of 38 to 74mm.

The piles were designed to be driven in four sections to reach a final vertical driven penetration depth of 116.6m. The hammers used to drive the piles for this platform were IHC S-600 and IHC S-1200 double acting hydraulic hammers.

The site-specific geotechnical borehole indicated that the subsurface conditions consisted predominantly of clays, with interbedded layers of dense sands. The soil profile is summarized in **Table 2.8.3-1.**

Sl. No	Depth, m	Soil Type
1	0-15	Clay
2	15-25	Sand
3	25-37	Clay
4	37-39.5	Sand
5	39.5-41	Silt
6	41-44.2	Sand
7	44.2-64	Clay
8	64-74	Sand
9	74-102	Clay
10	102-105	Sand
11	105-116.6	Clay
12	116.6-119.3	Sand
13	119.6-125.2	Clay

Table 2.8.3-1.: Summary of soil stratigraphy

All piles have been driven using IHC-600 and IHC S-1200 hammers.

However, pile B1 refused at 113m (3.6m short of design penetration) with blow counts per penetration exceeding 250bl/0.25m consecutively for six penetrations using IHC-1200 hammer with a hammer efficiency of about 80%.

Hammer malfunctioned when efficiency increased to 90%. The contactor was asked to resume driving after the hammer had been repaired in about 7hrs. However, pile moved only about 0.2m with a blow count of 1672/0.2m at 90% hammer efficiency. The hammer performance could not be verified as the piles were not instrumented. The client decided to conduct a restrike test after about 3.8 days with pile instrumentation so that the hammer performance and the available pile capacity could be verified. A total of 100 blows were given during the restrike with hammer efficiency ranging from 70 to 97%.

It was intended to estimate the available pile capacities during the continuous driving and the restrike test. As the piles were not instrumented during continuous driving, the hand cast analysis based on the field blow counts and hammer energy readings from console was used to estimate the pile capacity. Whereas, for restrike test, the available pile capacity was estimated by CAPWAP analysis on the measured force-velocity signature.

Wave Equation Analysis and Results:

The unit resistance values, both for shaft resistance and end bearing, from the API method were entered into GRLWEAP software for analysis and then iteratively modified to match the blow counts obtained from all the piles. Matching the blow count from the end of driving or when a significant driving interruption occurs are the most critical factors for calculating the capacity of piles; therefore, selection of a representative driving record can significantly affect the computed pile capacity as well.

Figure 2.8.3-1 represents the recorded blow count of all three piles (A1, A2 and B2) driven to final penetration with the S-600 and S1200 hammers, as well as the average blow counts over the pile penetration and the WEAP match. **Figure 2.8.3-2** represents the hindcast match from WEAP for pile B1 only. Suitable modifications were made to the provided static analysis results to improve the blow count match by adjusting the unit shaft resistance values for clay layers at 44.2m from 60kPa to 120kPa and at 74m from 100kPa to 180kPa and the unit end bearing values were increased from 5.5 to 8MPa for the sand layer at 64m.

The soil dynamic parameters are summarized in **Table 2.8.3-2** below,

Shaft Quake	2.5mm – All soil types
Toe Quake	D/120 (18mm) for sand, D/240 (9mm) for very dense sand and D/60 (36mm) for clay
Shaft Damping	0.16 s/m for sand 0.65 s/m for clay
Toe Damping	0.5 s/m – all soil types

Table 2.8.3-2: Summary of soil dynamic parameters

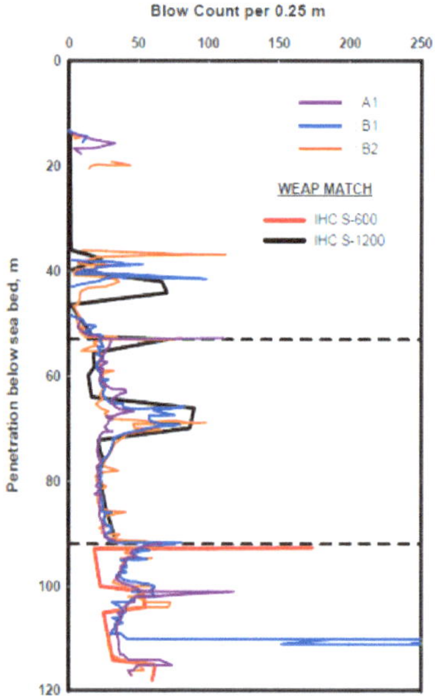

Figure 2.8.3-1: Hindcast match from WEAP for A1, B1 & B2 Pile

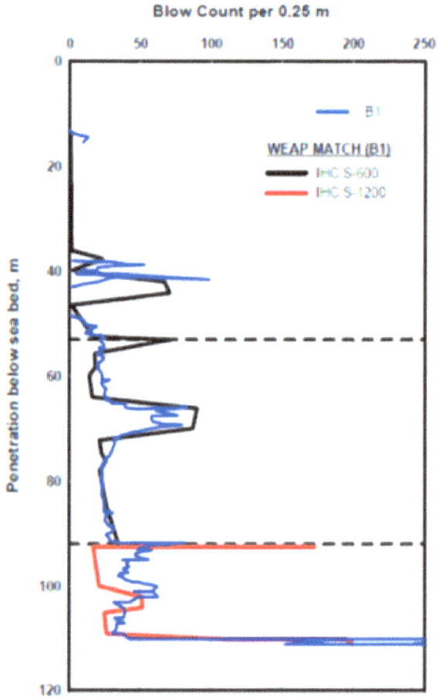

Figure 2.8.3-2: Hindcast match from WEAP for B1 Pile Restrike

CAPWAP:

The CAPWAP program performs a signal matching or reverse analysis based on measurements taken on a deep foundation under an impact load. The program is based on a one-dimensional mathematical model.

CAPWAP analysis was performed on the restrike data of Pile B1 to estimate the skin and toe resistance components. It is to be noted that the estimated capacity for refused pile is likely a lower bound value, and the true geotechnical value is likely higher because of insufficiency of the hammer to fully mobilize the geotechnical resistance.

Figure 2.8.3-3 shows the results of CAPWAP analysis. The analysis indicated a total capacity of 56.5MN with 47.2MN of shaft resistance and 9.3MN of end bearing.

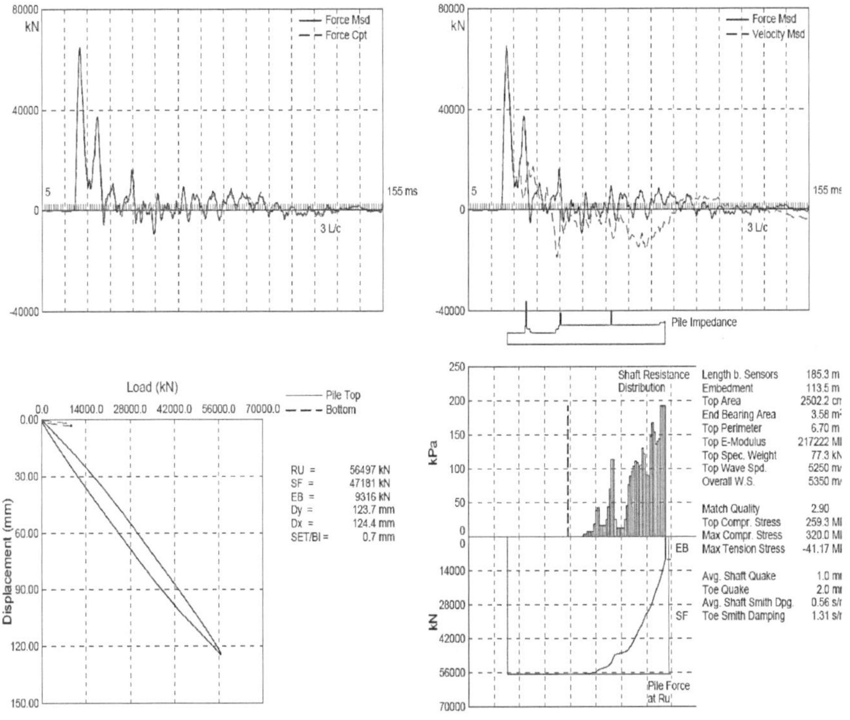

Figure 2.8.3-3: CAPWAP analysis results – B1 Pile Restrike

Conclusions:

The shaft resistance profile from the matching procedure was largely consistent with or higher than API calculated values, and, for evaluating the long-term capacity from the CAPWAP estimates and the hindcast study together, the mobilized end bearing at the refusal penetration of pile B1 was evaluated. The mobilized end bearing from the hindcast analysis initial drive (GRLWEAP) of pile B1 was 22.5MN. Adding this mobilized end bearing to the 47.18MN mobilized skin resistance from the restrike CAPWAP yielded 69.7MN. The total capacity thus estimated was more than the design compression pile capacity and this pile was accepted by the client. Wave equation analysis serves as a valuable tool for estimating the pile capacity based on the field pile behavior. For cases where refusal is encountered during restrike test, the fully mobilized end bearing from continuous initial driving can be superimposed to the mobilized shaft resistance of restrike to obtain a more realistic total capacity although it is still a lower bound estimate.

2.9 Associated Risks and Engineering Solutions

2.9.1 East Coast

Offshore in the eastern region of India, the geological makeup comprises predominantly of very soft to firm cohesive deposits extending to depths of 30 to 40 meters below the mudline. These deposits typically exhibit shear strengths of less than 50 kPa and are characterized by being heavily under-consolidated, with an over-consolidation ratio (OCR) of less than or equal to 0.5. This under-consolidation presents a significant challenge during pile driving operations, as negative pore water pressure may be induced within the soil strata. Additionally, the presence of alternating layers of sand or cohesionless material within thick cohesive layers is common in this area.

The under-consolidated soft cohesive deposits resting on the mudline pose a risk of excessive tilting of mud mats, necessitating careful consideration in the design process to determine the bearing capacity of the mud mat. Based on extensive experience, settlements of mud mats in this region are estimated to range from 400 to 800 mm below the mudline, a parameter crucial for ensuring the stability and performance of offshore structures. Being sedimentary in nature, the lateral variability of the strata is observed across different locations to a considerable extent. This variability underscores the importance of thorough site investigations and geotechnical/geophysical surveys to accurately characterize the subsurface conditions and inform the design and construction processes.

Proper understanding of lateral variability enables engineers to anticipate and mitigate potential geotechnical challenges, ensuring the safe and successful execution of offshore projects in the region.

In summary, the offshore region off the east coast of India presents unique geotechnical challenges characterized by under-consolidated cohesive deposits and alternating layers of sand. Careful consideration of these factors is essential in the design and execution of offshore structures and installations to mitigate risks such as excessive tilting of mud mats and thixotropic induced instability during pile driving operations. Thorough site investigations and accurate settlement estimations are crucial for ensuring the stability and performance of offshore infrastructure in this dynamic geological environment.

2.9.2 West Coast

The offshore engineering industry on the western coast of India faces numerous geotechnical challenges that require careful consideration during planning, design, and execution. This region features a diverse geological landscape comprising over-consolidated deposits, cohesive materials, and unpredictable cemented soil layers. Understanding these complexities is crucial for ensuring the safety, stability, and effectiveness of offshore structures, pipelines, and other infrastructure. This section delves into the geotechnical intricacies prevalent in the region and highlights the key design considerations required to address them effectively.

The western coast of India features a wide range of over-consolidated deposits with shear strengths typically falling within the range of 200 to 500 kPa. These deposits pose inherent risks due to their complex composition and varying mechanical properties. Deep-seated cohesive materials further exacerbate these challenges, making it essential to employ comprehensive site surveys using advanced geophysical and geotechnical methods. Despite intensive surveys, the presence of unpredictable cemented soil layers at considerable depths remains a common occurrence.

These layers often lead to premature refusal during pile driving operations, necessitating real-time monitoring to control stresses and ensure the attainment of the required pile capacity.

Cemented carbonate materials are frequently encountered at the mudline, presenting additional complexities in design and execution (viz., Pile self-weight penetration and associated stick up risk, Mud mat resting off the seabed due to non-penetration of jacket leg extension) Unlike regions with more homogeneous soil compositions, the western coast of India requires distinct design profiles for pile drivability and pile capacity estimation. The consideration of cemented carbonate materials as cohesionless deposits in capacity estimation ensures conservative design approaches, whereas accounting for their actual properties in drivability studies is essential for accurate assessments, of course, considering the effects of carbonate content. Designing offshore structures and pipelines in the presence of such geotechnical challenges requires a holistic approach that integrates geological, geophysical, and geotechnical data. Thorough investigation and characterization of the ground model are fundamental to understanding the site-specific conditions and variability.

Zoning the area based on soil properties and lateral variability further enhances the accuracy of design parameters.

During piling operations, the estimation of self-weight penetrations is critical to prevent engineering challenges during installation. Overestimation of penetration depths due to the consideration of cemented carbonate layers as cohesionless deposits can lead to excessive stick-up lengths posing difficulty for hammer placement. Realistic predictions of self-weight penetrations are imperative to mitigate such risks and ensure the structural integrity of pile-driven foundations.

Therefore, it is a common practice in the region to arrive at two different design profiles, one for pile capacity estimation where the cemented layers are considered as cohesionless deposits and the other for pile drivability where the cemented layers are considered as is. Both approaches being conservative estimates for their respective applications.

In pipeline design, the top 1m of soil plays a pivotal role in determining settlement patterns and stability. Differentiating between cemented and non-cemented layers at the mudline is essential for assessing settlement expectations and determining coating thickness requirements etc., during on bottom stability analysis. Furthermore, the presence of weaker materials necessitates careful evaluation of bearing capacity parameters to mitigate settlement-related issues and ensure long-term integrity.

Jack up leg penetration presents unique challenges, with soil conditions dictating the depth of leg penetration. Cohesionless materials and shallow cemented layers allow for shallow penetration, whereas cohesive materials permit deeper penetration. The risk of punch-through, particularly in cases of alternating soil profiles, underscores the importance of thorough site characterization and geotechnical analysis to prevent structural failures and operational hazards. The western coast of India presents a myriad of geotechnical challenges that demand meticulous attention and innovative solutions in offshore engineering projects. By integrating advanced site surveys, developing accurate ground models, and adopting appropriate design methodologies, engineers can effectively navigate these challenges and ensure the safety, stability, and longevity of offshore structures and pipelines. A proactive approach to geotechnical analysis and design is paramount for mitigating risks and maximizing the success of offshore projects in this dynamic and complex environment.

2.10 Lessons from the Experienced

2.10.1 From Naval Architect to Oil & Gas Leadership: A Journey through Infrastructure Development – V. Ashwini Kumar

Synopsis

The interview provides valuable insights into the evolution of India's infrastructure sector and the critical factors for success in large-scale projects. The interviewee's journey from naval architect to leading complex oil and gas projects demonstrates the importance of adaptability, thorough preparation, and maintaining high standards in project execution. As the industry shifts towards renewable energy and new geographical markets,

these fundamental principles of thorough investigation, quality maintenance, and schedule leadership remain crucial for success.

Introduction

In this insightful interview, a veteran industry leader shares his remarkable journey from naval architecture to becoming a prominent figure in oil and gas infrastructure development. With over four decades of experience spanning multiple continents and challenging projects, his story offers valuable lessons on project management, technological adaptation, and the critical importance of understanding ground conditions in infrastructure development.

Interviewee Profile

The interviewee is the Director, Oil & Gas BU at Afcons Infrastructure Ltd, heading their oil and gas division. Beginning his career as a naval architect and ship design engineer at Mazagon Dock, he later joined Larsen & Toubro in 1989, where he spent 25 years leading various offshore and onshore projects. His expertise spans across multiple geographies including India, Qatar, Abu Dhabi, Myanmar, and Tanzania. Prior to joining Afcons, he served as CEO of a joint venture with Sapura of Malaysia and headed onshore projects at IOT (Indian Oil Tanking). His leadership has been instrumental in executing some of the most challenging infrastructure projects in the region.

The Talk

CRP: Dr. C R Parthasarathy

AK: V Ashwini kumar, Director Oil & Gas BU, Afcons Infrastructure Limited, India.

CRP: Could you share your career journey in the infrastructure sector?

AK: I started as a naval architect and ship design engineer at Mazagaon Dock, then moved to Larsen & Toubro in 1989, marking my entry into oil and gas. We secured our first contracts with ONGC in 1991, and I was part of the startup team. Over 25 years at L&T, we executed projects across India, Qatar, Abu Dhabi, Myanmar, and Tanzania.

In 2005-2006, we ventured into offshore construction through a joint venture with Sapura of Malaysia, where I served as CEO. Later, I moved to IOT Indian Oil Tanking before joining Afcons about seven years ago to head their oil and gas division.

CRP: What were the defining moments in your career?

AK: The early 1990s projects for ONGC were significant turning points. We successfully executed large lumpsum turnkey (LSTK) projects that previously only Korean contractors like Hyundai could manage. Despite challenges, including the Surat plague that displaced three million people, we completed these projects on schedule, proving that Indian contractors could deliver complex LSTK projects efficiently.

CRP: How important is understanding the stratigraphy for oil and gas construction?

AK: Understanding stratigraphy isn't just important, **it's *critical*.** Let me give you an example. When we first worked in the Western Offshore, we'd do thorough geotechnical investigations, design piles, select hammers, and plan everything based on the data. But even then, we would hit surprises. Imagine a jacket 70–80 meters wide: one side of the pile would drive smoothly, and just 50 meters away, the other side would hit refusal. That's the reality of working with subsurface conditions—you have to be prepared for the unexpected.

In Abu Dhabi, we encountered solid coral and rock. Driving piles? Its Impossible. So we shifted to drilled-and-socketed piles, ie., you drill a borehole into the rock, insert the steel pile, grout it, and it's set - then no issues. But that required knowing the ground conditions upfront and having the right equipment ready.

Then there's IOCL Barauni, an onshore project. In one area, we used driven piles where the soil was stable and finished quickly. In another, we switched to bored piles—some 18 to19 meters deep—because the ground was unpredictable or required longer lengths.

Recently, on the jacket installation in East coast of India, we faced another challenge: as piles were near final penetration, the soil softened alarmingly and with blow counts dropping unexpectedly. The soil exhibited *thixotropic* behavior—losing strength during disturbance but regaining it over time. We had to pause, let the soil stabilize for months, and only then proceed with loading. It was a new lesson: even with experience, soil can behave unpredictably, and construction programs must account for such delays.

CRP: What are Afcons' key principles for maintaining industry leadership?

AK: Schedule leadership is paramount in EPC jobs. Completing projects on time typically results in satisfied customers and profitable outcomes. While we never compromise on quality, whether in concrete work or welding, ensuring timely material availability and proper resource allocation is crucial. Sometimes, additional resource investment for timely completion yields significant long-term benefits.

CRP: What significant challenges have you faced and what lessons were learned?

AK: One notable challenge was in Africa during a buried pipeline project. Insufficient geotechnical investigation led to unexpected clay conditions that our equipment couldn't handle effectively. We had to implement alternative solutions like concrete mattresses, which worked well but at additional cost. You can't control what's underground, but you *can* prepare. Have backup plans, contingency time, and flexible solutions.

No matter how thorough your geotech surveys, you can't test every inch of soil where foundations go. Strategic sampling helps us make educated assumptions about ground conditions, but surprises are inevitable—like hitting refusal in one spot and smooth driving in another.

That's why contingency plans are essential. Minimizing investment in upfront surveys might save time or money initially, but it'll cost far more to fix issues mid-construction.

Investing in solid geotech work and flexible engineering solutions upfront isn't just smart, it's what keeps projects on track when the ground throws you a curveball.

CRP: Who have been your key mentors or role models, and how have they influenced your career?

AK: Early in my career at Mazagaon Dock, senior leaders shaped my approach to people management and technical excellence. They emphasized empathy in leadership and precision in execution. Later, I learned equally from younger colleagues—one boss, 20 years my junior, challenged me to rethink traditional methods, highlighting the value of fresh perspectives.

CRP: What's your outlook for India's market in the next 3–5 years, and how are you positioning for growth?

AK: India's infrastructure sector shows tremendous potential despite recent development. While conventional offshore oil and gas opportunities may be plateauing, there's significant growth potential in renewable energy, particularly wind and solar.

AK: The Middle East, especially Saudi Arabia and affiliated regions, offers substantial opportunities, though entering these mature markets requires meeting high benchmarks in design, engineering, and construction expertise. Our strategy prioritizes infrastructure expansion and adapting to evolving energy trends.

CRP: How do you balance the demands of your leadership role with personal interests and family life?

AK: My days start early—5:30 AM—with physical activity: swimming, running. Weekends often involve trekking in Maharashtra, which keeps me energized. For stress relief, I turn to classical music, sketching, or art; creativity helps me disconnect and stay grounded.

Family is central. My wife and I care for my parents, both in their 90s, who live with us. Though they're independent, spending time with them—despite occasional communication challenges—is key. My elder son, a former army officer, lives in the UK with his wife and two kids. He met his wife during a UN mission in Congo—a story of chance and destiny. My younger son and his wife work and live in Mumbai.

Between work, hobbies, and family, staying active and present keeps the balance right.

2.10.2 From Field Engineer to Country Manager: Leading Innovation in Geotechnical Investigation at Fugro India - Bibhu Nayak

Synopsis

The interview provides valuable insights into the evolution of geotechnical investigation in India and the critical role it plays in infrastructure development. The interviewee's journey from field engineer to country manager exemplifies the growth potential in the industry, while his emphasis on innovation, quality, and personal well-being offers important lessons for industry professionals.

As India continues its infrastructure boom, the importance of proper geotechnical investigation and quality construction practices becomes increasingly crucial.

Introduction

In this insightful interview, we explore the remarkable 27-year journey of a leader in geotechnical investigation and his vision for infrastructure development in India. The conversation delves into the critical importance of geotechnical studies in infrastructure projects, innovative approaches to challenging projects, and the future of the industry in India's rapidly growing economy.

Interviewee Profile

The interviewee serves as the Country Manager of FUGRO India, bringing over 27 years of experience in geotechnical investigation and infrastructure development. Starting as a field engineer, he rose through various positions including technical manager, commercial manager, and business development manager before assuming his current role six years ago. His expertise spans complex offshore projects, deep-water investigations, and innovative geotechnical solutions. Under his leadership, FUGRO India has grown from 50 to over 650 employees, establishing itself as a leading provider of geotechnical services in the region.

The Talk

CRP: Dr CR Parthasarathy

BN: Bibhu Nayak, Country Director, FUGRO India

CRP: Could you share your professional journey with FUGRO?

BN: I started my career shortly after university graduation and joined FUGRO through a joint venture in 1996. Beginning as a field engineer spending 200 days annually in the field, I progressed through various roles from technical manager to commercial manager, business development manager, director, and finally to country manager for the last six years.

CRP: What has been your most significant project experience?

BN: A notable project was the first-ever deep-water project for Reliance on India's East Coast. Initially planned with 100% overseas participation, we transformed it within six months to operate with 100% Indian manpower, technologies, and assets, including complex deep-water data processing. This challenging project, which included offshore work and pipe laying in a narrow river with tight tolerances, significantly boosted our team's confidence.

CRP: How important is geotechnical investigation for infrastructure projects?

BN: Geotechnical investigation, along with geophysical investigation, is crucial for any infrastructure project. It's essential to understand soil strength and parameters before construction to prevent project jeopardy. Unfortunately, many organizations don't prioritize these investigations, though they represent a small portion of the total project cost. This oversight often leads to significant project downtimes and lifecycle issues.

CRP: What is FUGRO's approach to staying ahead in the industry and its vision for the future?

BN: FUGRO, which began as a geotechnical company in Netherlands 62 years ago, maintains leadership through continuous innovation, particularly in offshore wind farm technology. We develop new technologies in partnership with academia, focusing on deep water soil investigations and optimal foundation solutions for large offshore wind turbines.

I believe in empowering my team to own our vision. We focus on three things: *prepare for tomorrow, do things right* (no shortcuts), and *invest in people.* Starting with 50 employees 27 years ago, we're now over 650—and the future's bright. The next-gen leaders here are ready to push boundaries. For me, success is seeing people grow and the organization thrive. With this team, the sky's the limit.

CRP: What key lessons have you learned from geotechnical challenges in your projects?

BN: In my experience, the biggest lesson is this: *Don't assume the ground behaves the same everywhere.* Early on, we learned the hard way that even a few hundred meters can change stratigraphy completely. For example, in Uttarakhand, inadequate geotechnical information led to major issues—fortunately, no lives were lost, but a costly reminder.

We train teams rigorously to avoid repeating errors and prioritize thorough investigations. Mother Earth is unpredictable—respect her complexity and study every site meticulously. That's how you de-risk projects.

CRP: Who is the someone who has really shaped your career or life—and what's a lesson they taught you that stuck?

BN: My dad was my first role model. Growing up in a small town, he pushed me to focus on academics but also taught me to stick to my ethics and explore the world fearlessly. Later, my uncle—a self-made entrepreneur—taught me to prioritize quality over quantity. He'd say, '*You might only win one job out of ten but do that one right. Your reputation will grow, and people will follow you.*' Both lessons still guide me: ethics and excellence aren't optional—they're the foundation.

CRP: What are your views on India's infrastructure outlook for the next 3-5 years?

BN: India is experiencing significant growth in infrastructure development, with extensive road construction, tunnel projects, and railway network upgrades. The government is investing heavily in quality infrastructure, recognizing the long-term economic benefits of quality construction. This growth trajectory is expected to continue for at least the next decade.

CRP: How do you maintain work-life balance in such a demanding role?

BN: I advocate spending at least one hour daily on personal well-being. As an avid cyclist, I cover 7,000-8,000 km annually and participate in endurance cycling events and marathons. Beyond fitness, I love travelling both for work and leisure—and diving into technical journals to stay curious about innovations, from electric vehicles to space tech. Lately, I've been exploring music again, aiming to pick up the keyboard after years of playing the harmonium. It's all about balancing movement, learning, and creativity to stay grounded.

CRP: To wrap up, could you share a bit about your family and how they support your journey?

BN: I'm blessed with a loving family—my wife and I recently celebrated twenty seven years of marriage. Our two daughters keep us energized: My eldest graduated in computer engineering from the U.S. and works at Nvidia, while my younger one is in 9th grade. They're my anchors, reminding me daily that success starts at home.

2.10.3 Expert Insights: Decades of Offshore Geotechnical Experience of ONGC – K.K. Das.

Introduction

This interview elucidates the indispensable role of offshore geotechnical investigations. Expert insights are presented herein regarding the evolution of said practice, the significance of soil analysis, salient challenges, and instructive industry lessons. Furthermore, a detailed examination is provided of how these investigations constitute a foundational element for offshore projects, exerting influence over diverse facets encompassing structural design, installation procedures, and the processes of risk assessment and mitigation.

Note about the Interviewee

The interviewee is K.K.Das, GM(C), Head of the Geotechnical Section at IPEOT, Panvel, ONGC. The Geotechnical section is a department within IPEOT that specializes in offshore soil investigation for Indian offshore projects for ONGC.

The Talk (In correspondence)

Question: Could you provide an overview of the history and evolution of your organization's involvement in offshore geotechnical investigations?

ONGC: We would like to clarify that IEOT (Institute of Engineering and Ocean Technology) has been merged with IOGPT (Institute of Oil & Gas and Production Technology) and the new entity is now known as IPEOT (Institute of Production Engineering and Ocean Technology) .The two earlier institutes which came into existence since 1983 now caters to the technological needs of both onshore and offshore activities as a single knowledge center with its domain expertise.

With respect to Geotechnical section - Since 1994, Geotechnical Section has taken over the offshore soil investigation work from foreign agencies (viz. Fugro, McClelland Engineers, Dames & Moore etc.) and since then, all the ONGC platforms of western and eastern Indian offshore are designed /installed based on our soil investigation reports only.

Question: How critical is geotechnical investigation, including lab testing, analysis, and interpretation, to the success of offshore projects?

ONGC: Knowledge of soil properties below seabed is very essential to properly and safely design and install offshore structures. A large part of the commercial and operational risk involved in these works relates to uncertainties about the properties of soils at the site. The values of the geotechnical properties of the soil should be known with sufficient accuracy to meet the engineering requirements of the structural design. This requires a site-specific investigation and interpretation of the soil properties by geotechnical engineers. Geotechnical investigation which includes offshore/onshore lab testing, soil stratification and assessment of soil design parameters, engineering analysis (viz. ultimate axial pile capacity, axial and lateral load-deformation data, mud mat bearing capacity analysis, jack-up leg penetration analysis etc.) is the starting point for any projects to begin with and rest of the project activities take off their shape based on the soil investigation report.

The soil investigation report forms an integral part of bid documents and is the basis for finalization of jacket pile configuration and further detailed engineering of any offshore LSTK project. Platforms are designed and installed based on the Soil Investigation Report.

Question: What are some of the significant challenges encountered in offshore soil investigations, and what key lessons have been learned from these experiences?

ONGC:

Offshore always throws surprises and challenges but it also gives us a lot of lessons and enriches our experiences to handle the complex issues that are likely to come in future with more confidence and clarity. Some challenges and lessons / tips are following:

1. In western Indian offshore, for a platform location, generally the depth of soil investigation is up to 120 m to 130 m below seafloor. Soil stratification and assigning of soil design parameters itself for such a huge 120 to 130 m depth is quite a big, complex and challenging job.
2. Proper interpretation of the soil properties of the top seafloor layer(s) for the correct estimation of mud mat bearing capacity (for

fixed offshore jacket) is very critical since any incorrect design on the basis of it may result in overturning of jacket with severe consequence for the operator with respect to time and cost.

3. Incorrect interpretation of soil design parameters may result in selection of a wrong pile driving hammer which may result in pile refusal with serious consequences.
4. The assigned soil design parameters should result in an accurate and realistic picture of the leg penetration of the jack-up spud can foundation. Jack-up leg penetration analysis should be performed with utmost care to avoid possibility of rapid leg penetration or punch through, if any.
5. Generally, it is observed that in western Indian offshore, the Carbonate content (%) which influences the shear behavior especially in sands is quite significant. It may vary from 20-30% to 90-100%. Along with this, variable degrees of cementation are also observed. Generally, soils having significant carbonate content are not considered as good foundation material due to the crushing and compressibility of the material.

   Limit unit end bearing and limit unit shaft friction for sands should be assigned with full caution, considering the % carbonate content, degree of cementation and inclusion of fine grained soils etc. so that the pile foundation may be designed and installed safely in the offshore field.
6. Pile length effect – For long piles, driven in Clays, there may be degradation in ultimate axial pile capacity due to – continued shearing of a particular soil horizon during pile driving, lateral movement of soil away from the pile during pile driving, progressive failure in the soil due to strength reduction with continued displacement etc. So, for very long piles such as used in western Indian offshore for jacket platforms, the pile length effect should be taken into consideration for estimation of ultimate axial pile capacity.

Question: Who are some of the key role models, inspirations, and mentors who have significantly influenced the development of your organization's expertise in offshore geotechnical investigations?

ONGC: Geotechnical Section witnessed a lot of hard working and dedicated executives since its inception.

So many names will appear as we look back but there are two officers whose names will always be discussed and remembered in Geotechnical Section. The two foundation pillars of Geotechnical Section are – Dr RK Ghanekar, former Head, Geotechnical (Retired) and Mr. Rupam Mahanta, former Head, Geotechnical, currently working in ONGC. The two officers put their tremendous efforts and brought Geotechnical Section from a preliminary level in 1994 up to this current level in 2024. They are inspirations and role models for everyone working in the Geotechnical Section.

Question: What is your vision and what are the future plans for your organization in the field of offshore geotechnical investigations?

ONGC: Vision and plan for future: To develop knowledge and gain experience in deep-water soil investigation and soil investigation for offshore wind farms.

❖

From page 45, The procedure to pull out jack-up spudcans depends on the extent of its penetration below the mudline. Jacking systems generally prove sufficient to pull out spudcans that have penetrated less than 10.0m (low penetrated spudcans). If significant resistance is encountered during pull-out, the hull can be further lowered below the normal floating draught position to facilitate pull out. The pull out force is estimated using the following expression as per Clarom.

$$\mathbf{Q_p = W_l + W_s + Q_s + A*q_s}$$

Where,

W_l = submerged weight of leg (MN)

W_s = submerged weight of sediments accumulated on spudcan (MN)

Q_s = vertical shear force of accumulated sediments (MN)

q_s = equivalent pressure to section (MN/m^2)

A = spudcan area in contact with soil (m^2)

3

INDONESIA

3.1 General Geology of the Region

Indonesia is an archipelagic country extending about 5,120 kilometers (3,181 mi) from east to west and 1,760 kilometers (1,094 mi) from north to south. It comprises five main islands: Sumatra, Java, Borneo (known as Kalimantan in Indonesia), Sulawesi, and New Guinea; two major island groups (Nusa Tenggara and the Maluku Islands) and sixty smaller island groups.

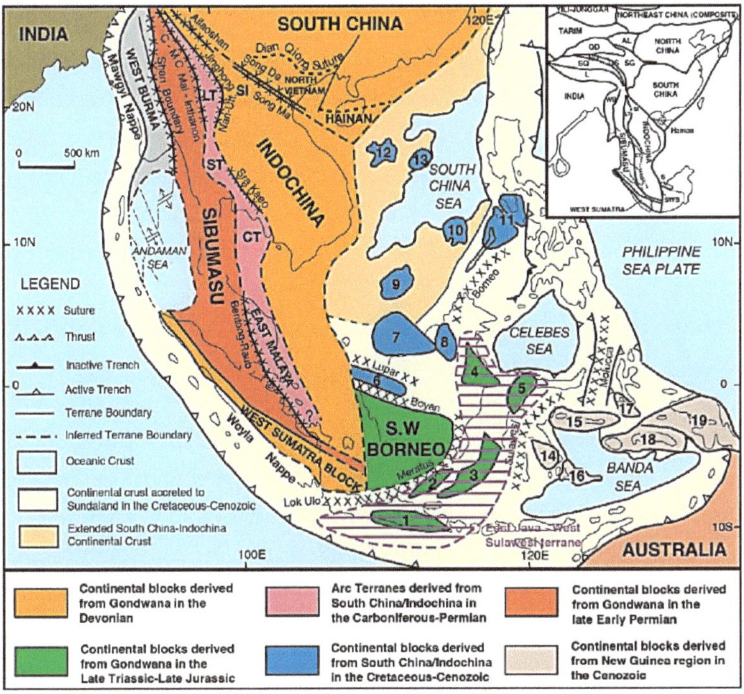

Figure 3.1-1: Southeast Asia- Indonesia Basement Terrain (Van Gorses 2018)

Indonesia is a transcontinental country, where its territory consisted of islands geologically considered as part of either Asia or Australia **Figure 3.1-1** shows general representation of basement terrains of Indonesia.

The tectonics processes in Indonesia formed major structures in Indonesia. The most prominent fault in the west of Indonesia is the Semangko Fault or the Great Sumatran Fault, a dextral strike-slip fault along Sumatra Island (about 1,900 km). The formation of this fault zone is related to the subduction zone in the west of Sumatra.

Palu-Koro fault is another major structural feature formed in the central part of Indonesia. This fault runs across the central part of Sulawesi Island and extends offshore to the west across Makassar Strait and ends in the Mangkalihat Peninsula in Borneo. The fault is named after the capital city of Central Sulawesi, Palu, on the west coast of Sulawesi and the Koro River, which is formed by the fault zone.

Sorong fault is a significant left lateral fault in the eastern part of Indonesia, named after Sorong City. It has east-west orientation and extends from the northern part of West Papua to East Sulawesi for about 2000 km. The regional condition of Indonesia which is surrounded by active plates produces a variety of geological conditions starting from Quaternary rocks, recent volcanic formations, Cenozoic formations, Mesozoic formations, Paleozoic formations, plutonic rocks, metamorphic rocks, and active volcanoes with a high potential for earthquakes and volcanic eruptions. For this reason, engineering geology considerations are very important in planning infrastructure development to produce infrastructure products that are faster, more efficient, of better quality, safer and more sustainable.

Offshore petroleum exploration in Indonesia began in late 1960's and thereafter a significant number of seismic data sets were acquired. Based on these data sets, 60 sedimentary basins were officially recognized by the Government of Indonesia (Sunarjanto et al., 2008). Understanding of those sedimentary basins is very important for petroleum exploration, as they are basically the places to find hydrocarbons. Petroleum potential within a basin is related to its sediment accumulation and tectonic history. Critical petroleum system elements such as source rocks, reservoir and seal mainly comprise sedimentary rocks. Indonesia is Southeast Asia's largest economy with a Gross Domestic Product of US$1.042 trillion in 2018, ranking 7th in the world based on Purchasing Power Parity.

Economic growth in 2019 decreased slightly to 5 %, compared to the previous year at 5.2%; growth over the last five years has been just above 5% per annum, credited to strong private consumption. In 2018 and 2019, investment in Indonesia's oil and gas industry reached around $12.3 billion and $12 billion respectively, decreasing from around $16 billion in 2016. Indonesia still has significant reserves of oil and gas but needs substantial investment and technology for the exploration. Much of the remaining reserves will require sophisticated technology that Indonesia's SOE operators do not have, and protectionist policies are discouraging global energy companies from pursuing opportunities, limiting the country's output. Overall, the country's oil production is declining by 11% per year due to ageing oil fields and the lack of new oil field exploration. In 2019, the country's oil production was 746,800 barrels per day (bpd), lower than 2017's production of 803,000 bpd. 2018 production decreased again with 778,000 bpd. Gas production for 2018 was 1.14 million barrels of oil equivalent per day (mboepd, Thousand Barrels of oil equivalent per day) which decreased slightly, 0.01 mboped from the previous year.

Most oil upstream activities are focused on western Indonesia, with the main areas for oil production being:

a) Sumatra
b) The Java Sea
c) East Kalimantan
d) the Natuna Sea
e) Matura Strait

The government has been encouraging exploration activities in eastern parts of Indonesia, where, according to the 2020 PwC Guide, 39 tertiary and pre-tertiary basins show rich promise in hydrocarbons.

Indonesia's main areas for gas production are:

a) South Sumatra
b) East Kalimantan
c) the Natuna Sea
d) Sulawesi
e) West Papua

The geological characteristics of Indonesian oil and gas fields are influenced by the dynamic tectonic processes in the region, giving rise to a variety of geological features that impact the accumulation of hydrocarbons and guide exploration strategies.

3.2 General Stratigraphies of Offshore Indonesia

The following section provides a basic understanding of the typical stratigraphy of some of the most important offshore oil and gas fields in Indonesia. (Refer to **Table 3.2.1-1, Table 3.2.2-1, Table 3.2.3-1 and Table 3.2.4-1).**

3.2.1 Madura Strait

Depth (m)	Description	Shear Strength (S_u), kPa		Angle of Internal Friction (phi)
		(Top)	(Bottom)	
0 – 20	Soft to Firm CLAY	2	20	-
20 – 25	Firm CLAY	35	50	-
25 – 30	Firm to Stiff CLAY	40	60	-
30 – 40	Firm to Stiff CLAY with Silt and Sand pockets	30	75	-
40 – 50	Stiff Calcareous CLAY	75	100	-
50 – 65	Stiff CLAY	75	100	-
65 – 75	Stiff to Very Stiff Calcareous Silty CLAY	90	120	-
75 – 100	Very Stiff Calcareous, shaly CLAY	155	175	-
100 – 105	Very Stiff Calcareous Shaly CLAY with sand partings	125	175	-
105 – 130	Very Stiff Calcareous Shaly CLAY with traces of shell fragments and organic materials	125	200	-

Table 3.2.1-1: Generalized Soil Stratigraphy of Madura Strait

3.2.2 West Java Sea

Depth (m)	Description	Shear Strength (Su), kPa		Shear Strength (Su), kPa
		(Top)	(Bottom)	
0 – 1	Firm to Stiff Clayey SILT	20	50	-
1 – 3	Loose to Medium Dense Clayey SAND	-	-	32
3 – 9	Loose to Medium Dense Silty SAND	-	-	32
9 – 12	Very Stiff Clayey SILT	100	150	-
12 – 15	Very Stiff Sandy CLAY	120	175	-
15 – 17	Very Stiff Sandy SILT	150	200	-
17 – 30	Very Stiff Silty CLAY	125	225	-
30 – 35	Very Stiff Clayey SILT	175	200	-
35 – 40	Loose to Medium Dense Silty SAND	-	-	32
40 – 50	Very Stiff Silty CLAY with Sand Partings	150	200	-
50 – 55	Loose to Medium Dense Silty SAND	-	-	32
55 – 65	Loose to Medium Dense Silty SAND with Clay seams and partings	-	-	33
65 – 85	Very Stiff Silty CLAY with sand pockets	125	200	-
85 – 87	Loose Silty SAND	-	-	30
87 – 90	Very Stiff Silty CLAY with Sand seams	125	200	-
90 – 95	Medium Dense Clayey SAND	-	-	32
95–100	Very Stiff Silty CLAY with sand laminations	125	200	-

Table 3.2.2-1 : Generalized Soil Stratigraphy of West Java Sea

3.2.3 East Kalimantan

Depth (m)	Description	Shear Strength (Su), kPa		Angle of Internal Friction (phi)
		(Top)	(Bottom)	
0 – 3	Very Soft Sandy CLAY	2	10	-
3 – 5	Soft to Firm CLAY	20	50	-
5 – 40	Firm to Stiff CLAY	40	75	-
40 – 60	Stiff Silty CLAY	50	100	-
60 – 65	Medium Dense SAND with clay seams	-	-	33
65 – 70	Very Stiff Silty CLAY	100	125	-
70 – 72	Loose Sand with CLAY seams	-	-	32
72 – 78	Very Stiff Silty CLAY	100	125	-
78 – 80	Dense Sand with CLAY seams	-	-	34

Table 3.2.3-1: Generalized Soil Stratigraphy of East Kalimantan

3.2.4 Natuna Sea

Depth (m)	Description	Shear Strength (S_u), kPa		Angle of Internal Friction (phi)
		(Top)	(Bottom)	
0 – 2	Very Soft Silty CLAY	2	10	-
2 – 12	Soft Silty CLAY	20	50	-
12 – 20	Very Stiff Silty CLAY	50	150	-
20 – 40	Stiff Silty CLAY with some partings of Silt	50	100	-
40 – 52	Stiff Sandy SILT with organic pockets	40	100	-
52 – 66	Very Stiff Silty CLAY with traces of silt	75	150	-

Depth (m)	Description	Shear Strength (S_u), kPa		Angle of Internal Friction (phi)
		(Top)	(Bottom)	
66 – 85	Stiff to very stiff Silty CLAY with silt partings	50	150	-
85 – 90	Stiff Clayey SILT	75	100	-
90 – 115	Very Stiff Silty CLAY with organic pockets	75	150	-
115-130	Very Stiff Silty CLAY with Silt pockets	125	175	-

Table 3.2.4-1: Generalized Soil Stratigraphy of Natuna Sea

Introduction

Offshore Indonesia is a region of remarkable geological complexity and geotechnical significance, positioned at the convergence of several tectonic plates within the Pacific Ring of Fire. This chapter explores Indonesia's offshore tectonic framework, sedimentary basins, soil characteristics, and engineering challenges. The information is tailored to offshore geotechnical engineers, providing insights essential for infrastructure development in this dynamic environment.

3.3 Geological and Geotechnical aspects of Offshore Indonesia

3.3.1 Geological Setting

Tectonic Framework

Indonesia lies at the intersection of the Indo-Australian Plate, Eurasian Plate, Pacific Plate, and Philippine Sea Plate. This tectonic interaction creates a highly active geological environment characterized by:

- **Subduction Zones**: The Sunda Trench and Java Trench are prominent features where the Indo-Australian Plate subducts beneath the Eurasian Plate, generating frequent seismic activity and influencing sediment deposition patterns.
- **Volcanic Arcs**: Active volcanic arcs contribute volcaniclastic sediments to offshore regions, affecting seabed composition.
- **Transform Faults**: Strike-slip faults, such as the Sumatra Fault System, create localized deformation zones that impact sediment stability.

Sedimentary Basins

Indonesia's offshore regions host numerous sedimentary basins with diverse depositional environments:

- **North Sumatra Basin**: Dominated by Paleogene and Neogene clastic sequences, offering significant hydrocarbon potential.
- **South Sumatra Basin**: Features interbedded sandstone and shale layers with prolific hydrocarbon reservoirs.
- **East Java Basin**: Comprises deltaic and deep-marine deposits influenced by fluvial systems from Java Island.
- **Banda Sea Basin**: Contains fine-grained pelagic sediments and occasional turbidite deposits in deep marine settings.

These basins exhibit lateral and vertical heterogeneity due to varying depositional processes, which directly influence geotechnical properties.

3.2.2 Geotechnical Characteristics

Soil Composition and Properties

The soils in offshore Indonesia vary significantly based on their origin and depositional environment:

- **Clayey Soils**: Found in deeper waters and back-arc basins; they are typically soft to medium-stiff with high plasticity. These soils are prone to consolidation settlement under loading conditions.
- **Sandy Soils**: Common in shallow shelf areas near river mouths; these are often poorly graded with variable density and susceptible to liquefaction during seismic events.
- **Volcaniclastic Deposits**: Composed of ash, pumice, and pyroclastic materials; their strength and compressibility depend on weathering and cementation levels.

Geohazards

Key geotechnical challenges arise from Indonesia's active tectonic setting:

1. **Seismic Activity**: Frequent earthquakes necessitate foundation designs that can withstand dynamic loads.
2. **Submarine Landslides**: Steep continental margins pose risks to pipelines and other infrastructures.
3. **High Pore Pressures**: Over pressured zones in sedimentary basins require careful management during drilling operations.
4. **Corrosive Environments**: High salinity accelerates corrosion of steel structures, demanding robust protective measures.

3.4 Case Studies

3.4.1 Case Study – 1 (Dr Prashanth Talkad and Dr CR Parthasarathy)

Introduction

West Madura Offshore (WMO) is situated in the offshore waters to the west of Madura Island, part of Indonesia's East Java province. This area is significant for its oil and gas reserves, making it a vital hub for energy exploration and production activities.

West Madura Offshore plays a crucial role in Indonesia's energy sector, contributing to the country's domestic energy supply and revenue generation through oil and gas exports. This case study is referenced to Madura offshore.

Problem Statement

The soil, in the top 10m (or up to leg penetrations), is disturbed by the Jack-up leg penetrations and extractions, consequently, the stresses induced by shallow foundations such as Jacket mud mat is to be critically assessed. The disturbed soil is a potential hazard, as it might affect the stability of the future installations nearby.

This case study focuses on evaluating the strength of the disturbed soil and examining its impact on the on-bottom stability of a Jacket structure, which was set to be installed next to the depression area created by the movement of the Jack-up rig, as well as in relation to the existing jacket structure.

The complex soil-structure interaction was modeled in Plaxis-3D and the critical V-H-M load combination for mud mat settlement was determined. The topsoil movement towards the Jack-up depression and the subsequent mud mat settlement and tilt were determined.

Project Details:

The geophysical seabed survey performed at an offshore location in West Madura, Indonesia, indicated multiple seabed hazards, including the Jack-up footprints from the previous rig move and the existing Jacket platform. A jacket platform was planned to be installed next to the footprints. The Jacket platform is temporarily supported initially by the mud mat before driving the piles.

The mud mat is subjected to combined vertical, horizontal, and moment (VHM) loading due to eccentric gravity load, piling sequence, and environmental actions.

The remolded soils in the seabed depressions close to the mud mat will affect the on-bottom stability of the Jacket during installation. The two jack-up rig footprints (viz., footprint 2 and footprint 5) of diameters ranging from 24m and 27m and the existing platform footprint of diameter 18m would potentially affect the stability of the mud mat. The surface of the two Jack-up footprints was found depressed by 3m compared to the natural seabed surface.

The summary of seabed hazards is provided in **Table 3.4.1-1**. The mud mat geometry of the proposed jacket is shown in **Figure 3.4.1-1**.

As the problem is three-dimensional and cannot be analyzed with simplified approaches and/or two-dimensional analysis, Finite Element Analyses were undertaken using PLAXIS 3D software.

Sl. No.	Type of Disturbance on Mud-line	Size in Diameter (m) (approximate)	Northing (m)	Easting (m)
1	Jackup rig Footprint 5	27.12	9271784.85	701461.39
2	Jackup rig Footprint 2	24.18	9271807.2	701498.36
3	J U FP-5	18.00	9271822.36	701463.01

Table 3.4.1-1: Summary of Seabed Hazards

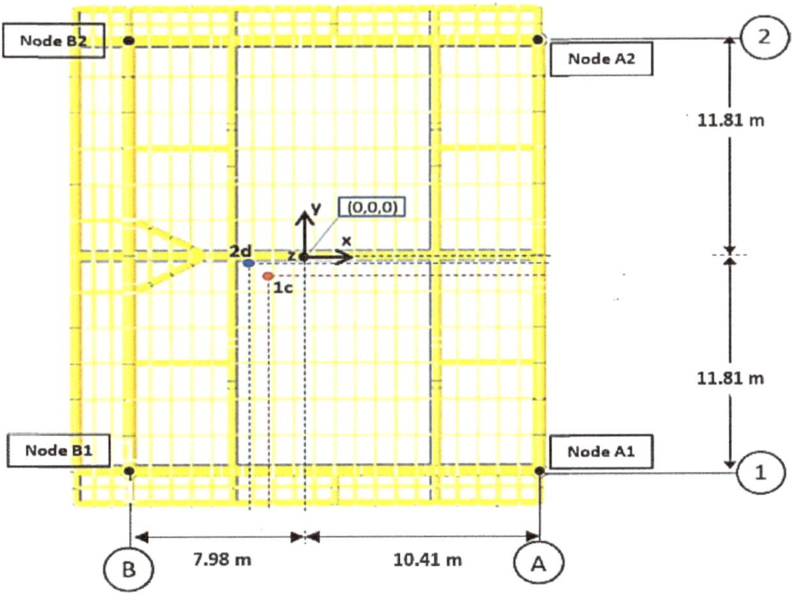

Figure 3.4.1-1: Mudmat geometry of the proposed Jacket

The geotechnical investigation consisted of one borehole with sampling up to a depth of 140m and one bump-over borehole with PCPT up to 140m. The seabed consisted of Very Soft Clay up to 2.8m followed by Firm Clay up to 7m. The soil stratigraphy predominantly consisted of cohesive soils with consistency ranging from stiff to very stiff at deeper depths. A Jack up rig installed at Clay location can create several meters of deep footprints after the legs are retracted.

In the case of very soft Clay, the Jack up spud can may have penetrated much deeper than the observed footprints. When the penetrated spud can is retracted, much of the soft remolded Clay will flow back and close the hole created by the leg penetration. The soil from backflow is disturbed and has a lower strength and stiffness than the surrounding intact soil.

The footprints and the associated remolded soil can be a potential hazard for new Jacket installation, particularly on the on-bottom stability of the mud mats. Therefore, to evaluate the hazard, separate soil design profiles were considered for jack-up footprints with remolding effect and the existing platform footprint and the surrounding intact soil as summarized in **Table 3.4.1-2 and Table 3.4.1-3.**

These design profiles were developed based on the thixotropic effect measured by performing UU tests on remolded samples over the thixotropic time intervals of 0, 3, 7, 14, and 30 days.

Footprint Type	Depth (m)	Soil Type	Shear Strength (Su), kPa	Unit Weight (γ) (kN/m3)	Poisson's Ratio (μ)	Modulus of Elasticity (E) (kPa)
Jackup Rig Footprints 2 & 5	0.0 – 3.0	Soil Removed due to Jackup Footprint				
	3.0 – 7.0	Firm Clay	20	15	0.49	6000
	7.0 – 10.0	Firm Clay	32	16	0.49	9600
Existing Jacket	0.0 – 3.0	Very Soft Clay	2.25	14.8	0.49	675
	3.0 – 7.0	Very Soft Clay	26	15	0.49	7800
	7.0 – 10.0	Very Soft Clay	37.5	15	0.49	11250

Table 3.4.1-2: Soil design parameters for Jack up footprint and Jacket footprint

Depth (m)	Soil Type	Shear Strength (Su), kPa	Angle of Internal Friction (phi), deg	Unit Weight (γ) (kN/m3)	Poisson's Ratio (μ)	Modulus of Elasticity (E) (kPa)
0 – 2.8	Very Soft Clay	2.5	-	14.8	0.49	750
2.8 – 7.0	Firm Clay	35	-	16.8	0.49	10500
7.0 – 10.0	Stiff Clay	50	-	17	0.45	15000
10.0 – 15.0	Medium Dense Sand	-	25	17	0.45	20000
15.0 – 21.0	Firm Clay	35	-	17	0.45	10000
21.0 – 36.0	Stiff Clay	50	-	18	0.45	15000
36.0 – 42.0	Stiff Clay	60	-	18	0.42	18000
42.0 – 50.0	Very Stiff Clay	80	-	18	0.42	24000

Table 3.4.1-3: Soil design parameters for the Intact ground

Two load combinations comprising the weight of the jacket and pile installation sequence (as furnished by the client) are summarized in **Table 3.4.1-4.** Loads (Fx, Fy, and Fz) are applied to the nodes A1, A2, B1 and B2 in the FEM analysis.

Load at Nodes in (KN)											
A1			A2			B1			B2		
Fx	Fy	Fz	Fx	Fy	Fz	Fx	Fy	Fz	Fx	Fy	Fz
Case: 1C (Jacket weight + Pile Arrangement 1 and Hammer)											
-89.49	2.0	-593.83	-114.27	-50.35	-348.25	47.41	-73.97	-1220.59	50.17	21.17	-1088.47
Case: 2D (Jacket weight + Pile Arrangement 2 and Hammer)											
-89.85	7.25	-450.16	-97.90	-38.12	-517.06	45.77	-23.26	-890.08	51.59	25.44	-1008.32

Table 3.4.1-4: Load combinations used in FEM analysis

Plaxis-3D Modelling

The lateral dimensions of the 3D model are 80 x 80 m, while the model thickness is 50.0m with the soil stratigraphy described in **Table 3.4.1-2 and Table 3.4.1-3**. These dimensions were selected such that the model boundaries have negligible effects on the results. **Figure 3.4.1-2 and Figure 3.4.1-3** show a cylindrical zone of 27.12m, 24.18m, and 18.0 m diameter in the top Clay layers. These zones (the top 10m from the mud line) are fully remolded by the previous spud can/Jacket installation. In the area of footprint 5 and footprint 2, a seabed depression of 3 m is considered.

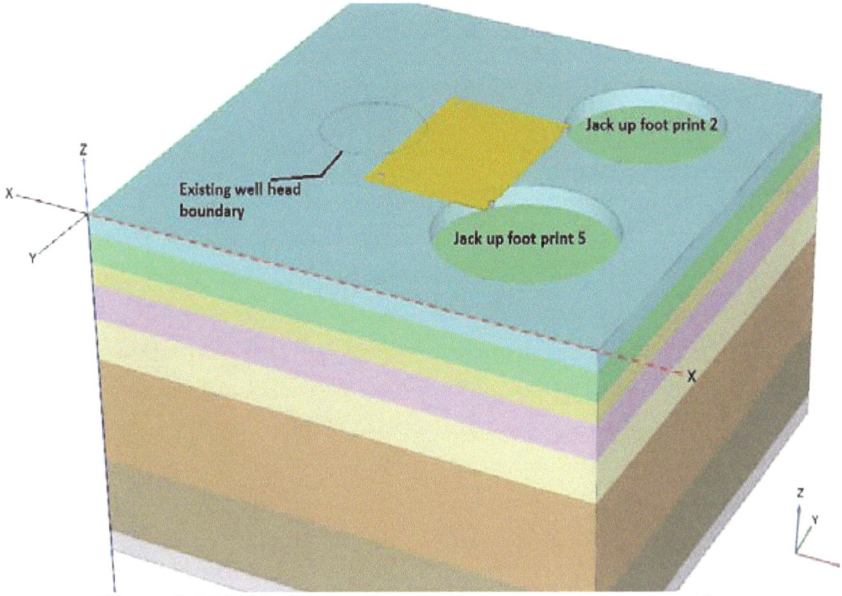

Figure 3.4.1.2: Model showing the footprint zones w.r.t mudmat

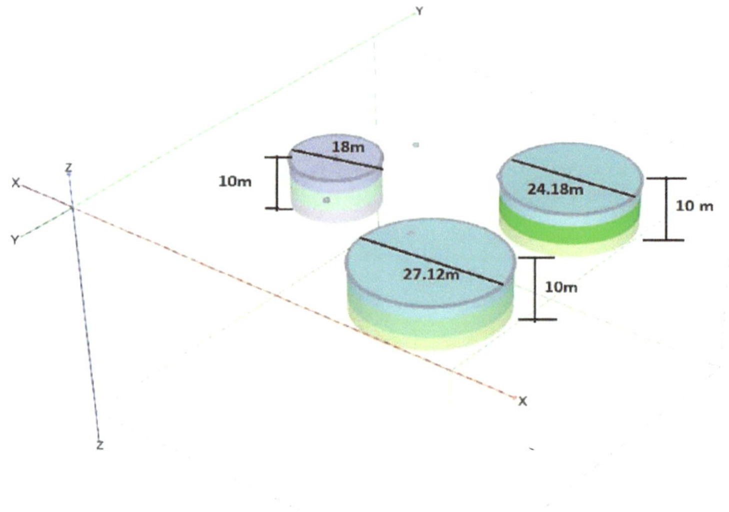

Figure 3.4.1-3 : Model showing the footprint zones with depth

The soil was modeled as an Isotropic Elasto-Perfectly plastic continuum, with failure described by the Mohr-Coulomb yield criterion.

The Clay layers are assumed to behave "undrained" and are characterized by a cohesion equal to the undrained shear strength, S_u. The elastic behavior was defined by a Poisson's ratio of 0.49 and a constant ratio of Young's modulus to undrained shear strength $E/S_u = 300$. The Sand layer is assumed to behave "drained" and is characterized by effective stress shear strength parameters c' and Φ'. The strength of the mud mat/Clay interface was modeled using an interface factor 'a', where the maximum shear stress at the interface is equal to the undrained shear strength.

The "rough" and "smooth" extremes of interface strength correspond to 1 and 0 respectively. An intermediate roughness was assumed as 0.6, which is a typical assumption for a Steel/Soft Clay interface. The Jacket mud mat is modeled as a rigid plain plate. The seabed is assumed to be perfectly flat below the mud mat. The mesh comprises several 10-noded tetrahedral elements. The mesh global coarseness was first set to "Medium", then the local refinement factor was decreased to 0.5 in the upper Clay layer, where the failure mechanism developed. The average element size was about 1m.

The Plaxis 3D analysis of the mud mat was carried out using the Staged Construction Technique by discretizing it into four phases as summarized below:

- Initial Phase: The initial Phase represents the soil layers with no influence of jack-up rig footprints.
- Phase 1: The jack-up rig footprints with crater marks are introduced.
- Phase 2: The mud mat and its interface are introduced.
- Phase 3: The loads acting at A1, A2, B1, and B2 are introduced.

The analysis was carried out for both the load cases as summarized in **Table 3.4.1-4** above.

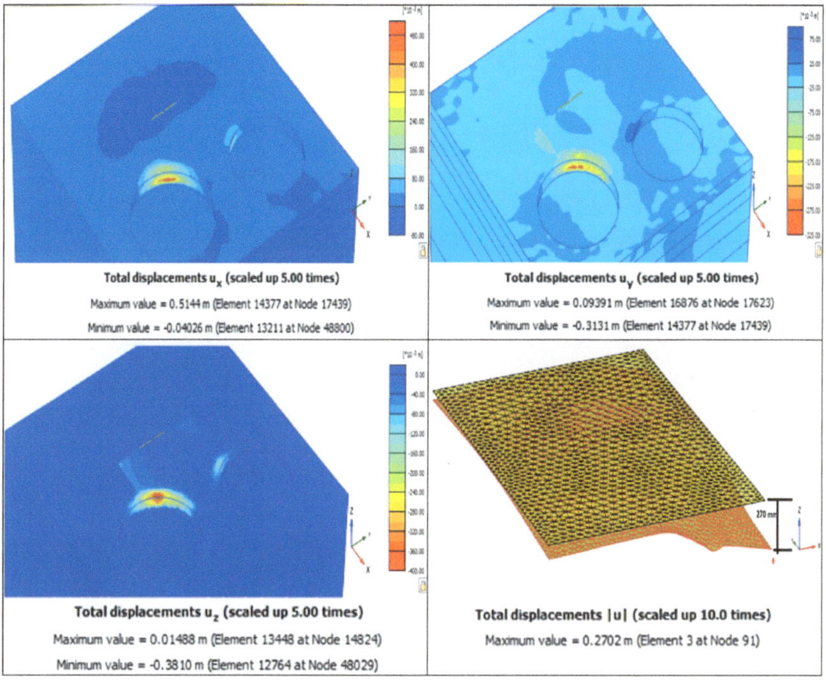

Figure 3.4.1-4: Total displacement in ux, uy, uz directions and maximum displacement at location A1 (Load case 1C)

Results

The finite element analysis of the mud mat provides the displacement contours of both the soil and the mud mats. The analysis for load case 1C had higher deformations due to higher loads compared to load case 2D.

The maximum vertical deformation of the mud mat occurred at the corner of the mat close to A1.

This is primarily because location A1 lies in proximity of the Jack up rig footprint 5. The deformation was around 270 mm at A1 location, while the tilt of the mud mat was around 0.46 deg. The edge of the crater (Jackup footprint 5) was expected to be under severe stressed condition during the installation of the mud mat. **Figure 3.4.1-4** provides the displacement contours and model deformation due to Load case 1C

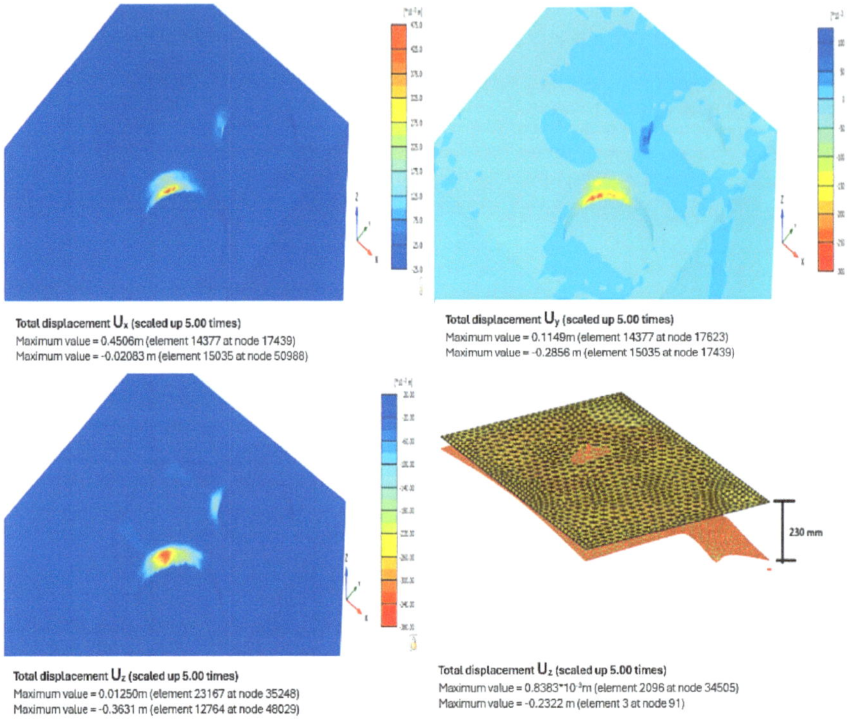

Figure 3.4.1.5: Total displacement in ux, uy, uz directions and maximum displacement at location A1 (Load case 2D)

The analysis for load case 2D resulted in a maximum displacement of 230 mm at A1. **Figure 3.4.1-5** shows the displacement contours and model deformation due to load case 2D.

The load combinations furnished for 1C & 2D were multiplied by a factor of 2.0 and Plaxis 3D analysis was carried out. The analysis resulted in no plastic failure up to a factor of 1.4 & 1.55 respectively for cases 1C & 2D for combined load application F_x, F_y, and F_z.

Conclusions

- A 3D finite element model was created in Plaxis software to study the behavior of mud mat with respect to the existing footprints within the vicinity.
- The Jack up rig footprints and existing platform location were modeled with a different shear strength profile compared to the original soil.
- Load case combination 1C (Jacket weight + Pile + hammer option 1) resulted in higher displacements of mud mat compared to the load case combination 2D.
- The maximum vertical displacements in both cases occurred at the corner of the mud mat close to the A1 location. This is primarily because of the influence of mud mat coming in contact with a large crater (Jack up rig footprint 5).
- The maximum vertical displacements were 270 mm and 230 mm for load cases 1C and 2D respectively.
- For load case 1C the maximum tilt is 0.46 deg.
- Out of two load cases, load case 1C seems to be critical for mud mat deformation. It was also verified that for the critical load case 1C, a factor of 1.4 was available.
- For both combinations of loads, it was evident that the soft Clay layer had squeezing effect (from the deformation plots). Any disturbance to the upper soil layers during installation will aggravate more deformation and tilt of the mud mats as the lateral flow of soil cannot be controlled.
- The relative shear stress (ratio of mobilized shear strength to max shear stress) was almost unity implying reaching a plastic state of soil for layer 1, particularly at the footprint locations.
- It was recommended to fill the craters (Jack up rig footprint area) with Sandbags before the installation of the Jacket to mitigate any additional hazards during the driving of piles

Based on the above analysis and recommendations the Jacket was installed without having any detrimental effect.

3.4.2 Case Study – 2 (Dr Ramesh Gangisetti, Dr CR Parthasarathy, Prof VS Chandrasekaran & Ms Nikita Keshav)

Introduction

In this case study, we explore the concept of conductor setting depth in a location situated in the East Java Sea, approximately 65km North of Madura Island. The conductor setting depth refers to the final depth up to which conductors are installed during the construction of offshore oil and gas drilling platforms. Conductors are small-diameter steel pipes that are driven into the seabed to provide structural support and to guide subsequent hydrocarbon drilling operations.

Technical Discussions

The scope of this case study was to review the available geotechnical information of a Jacket location and evaluate the conductor setting depth for the same. The soil substrata at this location consists of several layers of silty clay, a layer of silty sand and a layer of sandy silt. The silty clay layers vary in consistency from very soft to soft at shallower depths to stiff and very stiff at deeper depths. The sequence of layers from the mudline is as follows: Silty clay layers (0.0m -34.3m), dense sandy silt (34.3m-47.8m), stiff to very stiff silty clay layers (47.8 m-103.9m) dense silty sand layer (103.9m-107.4m) and very stiff silty clay layers (107.4m-130.4m). Based on the geotechnical data, the average submerged unit weights were estimated with respect to the depth, as shown in **Table 3.4.2-1.**

Depth range (m)	Average submerged unit weight (kN/m³)
0.0 to 103.9	7.66
103.9 to 130.4	8.00

Table 3.4.2-1: Average submerged unit weight

For the silty clay layers in the depth range from 0 to 80 m, Mesri (1975) equation $S_u/p'nc = 0.22$ is considered satisfactory. This corresponds to a PI value of 30 according to Skempton and Bjerrum correlation.

$$\frac{s_u}{p'_0} = \mathbf{0.11 + 0.0037}\, \boldsymbol{PI}$$ ***Equation 3.4.2-1***

The shear strength values obtained from various tests carried out (including those from cone penetration values) agreed well with the equation $S_u/p'nc = 0.22$ up to 80m depth. From 80 to 103.9 m the shear strength values correlated from CPT tests agreed with $S_u/p'nc = 0.22$ whereas those obtained from shear tests were slightly less. Further the pre-consolidation pressure p_c' determined at a depth of 88.4 m from laboratory oedometer test was 800 kPa against an estimated p_v' value of 677 kPa (7.66 x 88.4). This represented an OCR of 1.18. Additional Oedometer tests were assigned, and the results were re-analyzed to estimate pre-consolidation pressures as tabulated in **Table 3.4.2-2** below.

Sample	Depth (m)	p'v (kPa)	p'c (kPa)	OCR
P71	97.9	704.0	950.0	1.35
P72	93.9	719.0	930.0	1.29
P80	109.7	840.0	850.0	1.01

Table 3.4.2-2: Summary of design parameters

Based on the results, it was safely concluded that the silty clay layers were normally consolidated up to a depth 109 m depth. The PCPT also reconfirmed the same conclusion. There was a layer of silty sand from 103.9 to 107.4 m. This layer should have enabled 100% primary consolidation of adjoining silty clay layers. It was not advisable to have the conductor setting between 100 to 111 m to avoid development of seepage pressures in the silty sand layer. Beyond 107.4m the shear strength values are less than normal consolidation values as predicted by $su /p'nc = 0.22$. Based on the CPT values and the shear test results, a linear variation of shear strength from 160 kPa @ 107.4 m depth to 180 kPa at 130.4 m was adopted for this layer in the computation. The reduction in shear strength may be due to several reasons such as increase in silt percentage, reduced PI values and under consolidation. An Oedometer test carried out on a sample obtained from a depth of 115.8m gave a pre-consolidation pressure of 700 kPa against an overburden of 887.0 kPa (7.66 x 115.8). This indicated a consolidation ratio of about 79%. Lateral earth pressure coefficient K_o which depends on the over-consolidation ratio plays an important role in the computation of hydraulic fracture pressure. For normally consolidated clay, the following equations may be used:

$$K_o = 0.4 + 0.007PI \quad \textit{Equation 3.4.2-2}$$

$$K_o = 1 - \sin \varphi' \quad \textit{Equation 3.4.2-3}$$

Using the first equation **(3.4.2-2)** with a PI of 36, a value of Ko = 0.652 was obtained. According to the second equation this value corresponds to a φ' value of 20deg which was satisfactory. This value of K_o was used for silty clay layers up to 103.9m. A reduced value of K_0 = 0.6 was used for the clay layer below 107.4m to account for the possible under consolidation in this layer. The summary of design parameters used in this analysis is presented in **Table 3.4.2-3.**

Depth below Mudline (m)	Layer	Submerged Unit weight (kN/m^3)	Coefficient of earth pressure At rest K_o	Undrained shear strength kPa
47.8 to 103.9	Normally consolidated Silty clay	7.66	0.652	$0.22\,p'_v$
107.4 to 130.4	Silty clay	8.00	0.6	Linear variation 160 to 180

Table 3.4.2-3: Summary of design parameters

Conductor Setting Depth (ISO 3421:2022)

The setting depth of a conductor is decided on the following criteria:

1. The conductor should have adequate bearing capacity to carry safely the loads due to equipment mounted on it and the operational loads.
2. The soil below the conductor base should not undergo hydraulic fracturing due to the pressure of the drilling mud.

Conductor base will have to reach a depth below the mudline where the soil formation strength is sufficient to withstand the pressure from the drilling fluid. Otherwise, hydraulic fracture will result, and the drilling mud will flow into the formations which may even affect the integrity of the foundation. The analysis followed here was based on the approach proposed by Anderson and Lunne (1994). "A method to determine conductor setting depth in clay". Vol 1. Proceeding BOSS. According to this approach the conductor setting depth should be such that,

$$\mathbf{p_{max}} = \frac{\mathbf{p'_f}}{\gamma_1 \times \gamma_2} + \mathbf{u_0} \qquad \textit{Equation 3.4.2-4}$$

Where,

p_{max} = Maximum drilling mud pressure

p'_f = Hydraulic fracture pressure

γ_1 = Coefficient depending on the consequences of hydraulic fracturing.

γ_2 = Coefficient depending on the uncertainties in the soil data used to determine the hydraulic fracture pressure.

In normal cases γ_1=1 and γ_2 should vary between 1.1 and 1.3 depending on the quality of input parameters and the simplifications made in the calculation procedure. Drilling parameters are as shown below in **Table 3.4.2-4.**

Water Depth	: LAT EL (-)81.8m and HAT EL (-)83.4m
Mud Discharge Level	: EL (+)35.0m and EL (+)39.5m
Effective Unit Weight of Soil	: 7.66 kN/m³
Density of Drilling Mud	: 11.2kN/m³ (9.5ppg)
Cement Discharge Level	: EL (+)7.6m
Density of Cement	: 15.9kN/m³ (13.5ppg)
Safety Factor, r1	: 1.0
Safety Factor, r2	: 1.0, 1.15 and 1.2
Unit Weight of Sea Water	: 10.06kN/m³

Table 3.4.2-4: Drilling Parameters

Anderson and Lunne (1994) recommend the following three simplified procedures for estimating the hydraulic fracture pressure (p'_f).

Procedure 1

The hydraulic fracture pressure is estimated (without calculations) from a diagram which has been established from offshore in-situ hydraulic fracture pressure tests and laboratory tests on five different soils. This diagram is reproduced in **Figure 3.4.2-1**.

To use this method the over-consolidation ratio of the soil as well as the vertical effective stress at a given depth are required.

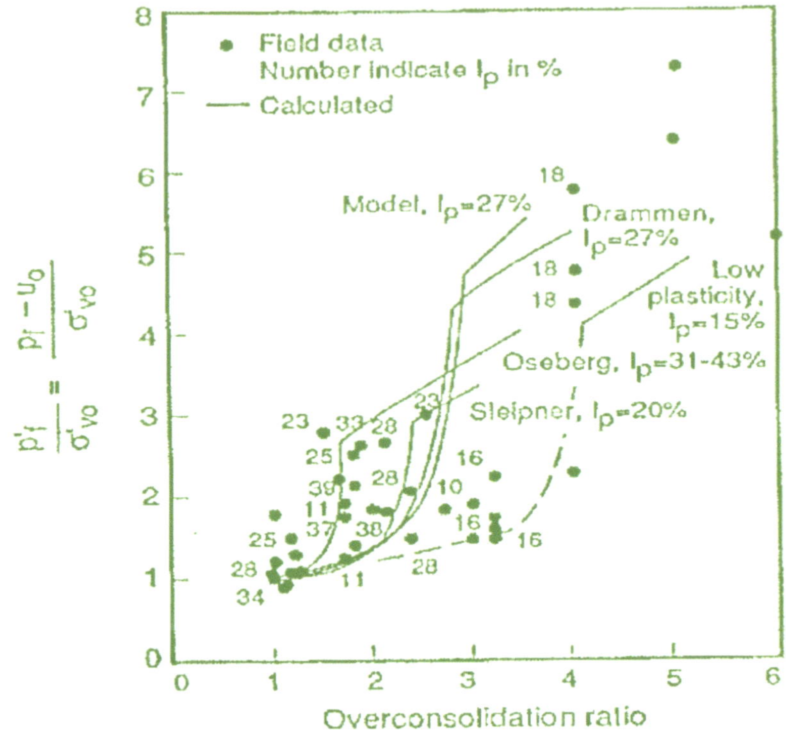

Figure 3.4.2-1: Calculated hydraulic fracture pressures for various soils based on DSS tests with shearing on the horizontal plane compared to the results of field tests (Anderson and Lunne 1994)

Procedure 2

In its simplest form as could be applied in the present case, hydraulic fracture pressure can be calculated using the below equation,

$$\frac{p'_f}{p'_v} = K_0 + 1.6\frac{s_u^{DSS}}{p'_v} = K_0 + 1.2\frac{s_u^c}{p'_v} \qquad \textit{Equation 3.4.2-5}$$

Where,

p'_f = hydraulic fracture pressure

K_0 =coefficient of earth pressure at rest

s_u^{DSS} =undrained shear strength obtained from the direct simple shear test

p'_v =initial vertical effective stress in the soil

s_u^c = undrained shear strength from anisotropically consolidated triaxial compression tests.

In the above equation s_u^{DSS} was approximately replaced by $0.75s_u^c$ by the authors. In offshore practice if s_u^{DSS} value is not available the values from other shear tests (UU) are used without any modifications.

Procedure 3

This requires laboratory undrained Direct Simple Shear (DSS_u) tests with pore pressure measurements on soil samples obtained at various depths. Since these results are not routinely available this procedure was not used in the present case. Only procedures 1 and 2 are used.

Results based on Procedure 2

The analysis was carried out for the following cases: Conductor discharge height above sea level of 35.0 m and 40.0 m. In all the analyses value of $\gamma1$ =1.0 was used. A range values of $\gamma2$ = 1.10, 1.15 and 1.20 were used. The conductor depth is arrived at as the intersection point of the mud pressure line and the p_{max} line which is presented in **Figure 3.4.2-2** below. The other cases are similarly plotted. The results are summarized below in **Table 3.4.2-5.**

Serial No	Mud discharge height (m)	γ_1	γ_2	Conductor Setting depth below mudline (m)
1	40	1.0	1.20	117.07
2		1.0	1.15	97.56
3		1.0	1.10	92.49
4	35	1.0	1.20	100.00
5		1.0	1.15	87.47
6		1.0	1.10	82.93
7*	**40**	**1.0**	**1.15**	**98.12**
8*	**35**	**1.0**	**1.15**	**87.98**
Note: *Procedure 1 is used with normal consolidation assumption				

Table 3.4.2-5: Conductor setting depth analysis results

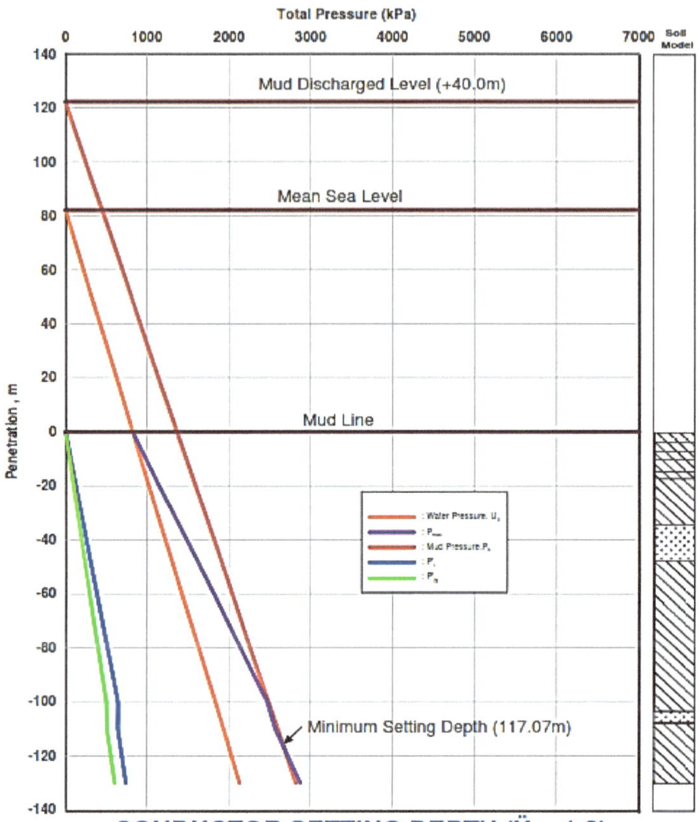

CONDUCTOR SETTING DEPTH (Ÿ$_2$= 1.2)

Figure 3.4.2-2: Conductor setting depth

Based on the analysis, a conductor setting depth of 117.07m below mudline was required if values of γ1 =1.0 and γ2 =1.2 are adopted (Procedure 2). Though the Oedometer tests shallower than 110m have indicated the soil to be normally consolidated, the Oedometer test carried out at the depth of 115.8 m indicated that the layer is under consolidated with a consolidation ratio of only 0.79. This implies there might be still some residual pore pressure to be dissipated. Even though too much emphasis on this one test may not be called for, it was better to search for better alternate setting depths. It was worthwhile to consider an alternative of using a setting depth of 97.56m with a reduced value of γ_2 =1.15. At this depth the soil is likely to be normally consolidated being closer to the drainage layer at 103.9 m.

Also, the result of the Oedometer test conducted at a depth of 88.4 m has shown that the soil at this depth was slightly over consolidated with an OCR of 1.18.

This was considered a better option even though a reduced value of γ_2 =1.15 was used. The thickness of the silty clay layer above the sand layer will be about 6m, which was adequate.

An additional analysis was carried out using procedure 1. This procedure required only the OCR of the deposit. Using the chart given in **Figure 3.4.2-1** and the assumption of normal consolidation with OCR =1.0 we get the ratio of $\frac{p'_f}{p'_v}$ = 1.0. The setting depth will then be 98.12 m using γ_2 =1.15 and mud discharge height of 40.0 m. This is shown in row seven of **Table 3.4.2.5** above. This value was in close agreement with the value of 97.56 m obtained by using procedure 2.

Conclusions

It was recommended to adopt a conductor setting depth of 97.0m below mudline which will have γ1 =1.0 and γ2 =1.15 for a mud discharge height of 40m. This also provided a cushion of about 7m above the silty sand layer.

The conductors were installed to 97.0m in this project and the wells were also successfully drilled using the Jack up next to the platform.

3.4.3 Case Study 3 (Dr Prashanth Talkad, Dr Gangisetti Ramesh, Ms Minnu KP, Ms Anusha Bekal & Dr CR Parthasarathy)

Introduction

Seawater intake structures are vital components of various industrial processes, particularly in sectors such as desalination plants, power generation facilities, and industrial manufacturing. These structures are designed to efficiently draw large volumes of seawater from the ocean for use in cooling systems, desalination processes, or other industrial applications. The design and construction of seawater intake structures are crucial for ensuring the reliability, efficiency, and environmental sustainability of the operations they support. These structures typically consist of intake pipes, screens, pumps, and sometimes additional filtration or pretreatment systems to remove debris, sediments, and marine organisms from the seawater before it enters the industrial process.

Problem Statement

A seawater head intake structure constructed in West Jawa of Indonesia was observed to have undergone excessive long-term settlement, potentially due to weak foundation soils. The foundation was laid on weak soil, which had been disturbed during dredging and further affected by the backflow of seabed mud. This case study deals with the assessment of the remolded strength of soils and the ground improvement technique adopted to mitigate the settlement. Ground Reinforcing Elements (GREs) were used to improve the stiffness of the ground and the soil-structure interaction with V-H-M loading was modeled in Plaxis-3D to estimate the settlement of intake structure with and without GREs. The results showed that GREs reduced the settlements significantly to values below the allowable limits.

Project Details

The footing of the intake head was designed to be embedded into a 1.2m thick concrete base resting on the ground at EL-9.403m, where the soil got remolded. Ground Reinforcing Elements (GREs) were proposed to improve the stiffness of remolded soil and the intact soils below. GRE is a reinforced concrete pile driven into the soil below the intake head footing, to improve the soil stiffness only but not to carry the structural loads.

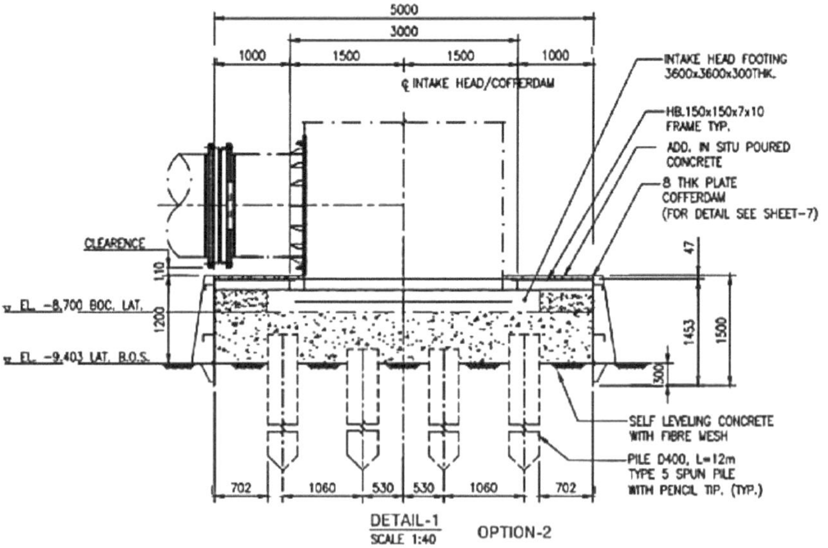

Figure 3.4.3-1: Cross-section of intake structure and the GREs

Sixteen Type-A spun piles of 12m long, 400mm OD were proposed to be driven within the concrete base footprint area.

The proposed GREs and the intake structure cross-section details of the intake head are presented in **Figure 3.4.3-1.**

The intake head installation procedure recommended is summarized below

a) Dredge out the topsoil to reach EL-9.403 LAT.

b) Drive 16 Nos of 12m long 400mm OD spun piles until the pile top of the peripheral row and internal row reach 0.5m and 0.3m, respectively, above the EL-9.403 LAT. Steel cofferdam is placed after the piling is completed.

c) Pour the self-levelling concrete into the cofferdam till EL-8.203 LAT.

d) Place the intake head and pour the concrete up to the brim of the cofferdam.

Technical Discussions

Based on the geotechnical investigation reports, the ground models were developed for

a) Unimproved ground and

b) Improved ground with GREs.

In the unimproved ground, it was assumed that dredging activity might remold the top 1m of stiff clay at EL-9.403m, and the backflow of mud gets mixed with the remolded soil. The dredging activity will affect the shear strength and consolidation parameters in the top 1m.

The depth of influence due to base pressure was considered up to 6m below the base. . The design parameters deduced from lab test results for the settlement assessment of the unimproved ground are presented in **Table 3.4.3-1.**

Layer	Depth below concrete pad, (m)	Description	C' (kPa)	Φ' (deg)	Saturated Unit Weight (kN/m³)	Initial Void ratio e_0	Compression Index Cc	Swelling Index Cs
1	0-1	Firm Clay	20	28	16	2.0	0.62	0.067
2	1-6	Stiff Clay	20	35	17.4	1.52	0.5	0.083

Table 3.4.3-1: Soil design parameters of unimproved ground

When the GREs are introduced, the soil will become stiffer, leading to reduced settlements. Additionally, as the piles are driven, the soil gets displaced and subjected to compression due to the adjacent piles' movement restraint. The stiffened ground will have improved shear strength and consolidation characteristics. The design parameters deduced from lab test results for the settlement assessment of the improved ground with GRE are presented in **Table 3.4.3-2.**

Layer	Depth (m)	Description	C' (kPa)	Φ' (deg)	Saturated Unit Weight (kN/m3)	Initial Void ratio e0	Compression Index Cc	Swelling Index Cs
1	0-20	Stiff Clay	25	35	19	1.52	0.25	0.05

Table 3.4.3-2: Soil design parameters of improved ground with GREs

Settlement Analysis

The foundation settlement analysis of an intake structure was evaluated using Plaxis 3D. The analysis was carried out considering the ground model without and with GRE as described below for pile grid optimization.

The loads provided as per the client,

Vertical Load = 639.68kN

Horizontal Load = 250.09kN

Moment = 1010.35 kN-m

Vertical Loading case:

- **Step1:** Evaluation of intake structure settlement without Ground Reinforced Elements (with only vertical load to calibrate the FEM results with analytical solution)
- **Step2:** Evaluation of intake structure settlement with Ground Reinforced Elements (For only Vertical Load case) - 16 numbers 12m long (4 x4 grid).

Vertical-Horizontal-Moment (VHM) loading case:

- **Step3:** Evaluation of intake structure settlement without Ground Reinforced Elements (For VHM Load case)
- **Step4:** Evaluation of intake structure settlement with Ground **Reinforced Elements (For VHM Load case)-16 numbers 12m long (4 x4 grid).**

- **Step (4 - A): Evaluation of intake structure settlement with optimized Ground** Reinforced Elements (For VHM Load case) - 16 numbers 8m long (4 x4 grid).
- **Step (4 - B):** Evaluation of intake structure settlement with optimized Ground Reinforced Elements (For VHM Load case) - 9 numbers 12m long (3 x3 grid).

The Concrete pad was modelled as plate element, with concrete properties. The thickness and dimensions were as per the geometry provided. The plan dimension of the concrete pad was **5.0m x 5.0m**.

PLAXIS 3D Model

The lateral dimensions of the 3D model are 20 x 20 m, while the model thickness was 20 m with the soil stratigraphy described in **Tables 3.4.3-1 & Table 3.4.3-2.** These dimensions were selected such that the model boundaries have negligible effects on the results. The soil was modelled with an advanced constitutive model available in Plaxis 3D to capture the long-term settlement characteristics. Isotropic hardening soil continuum was considered in the analysis for both the layers. The properties of the soil, in particular the compression index, swelling index and initial void ratio were considered based on the Oedometer data available with the nearest borehole. The strength of the Concrete PAD/Clay interface was modelled using an interface factor '1', treating as rough. The concrete pad was modelled as a rigid plain plate.

The surface below the concrete pad is assumed to be perfectly flat. The mesh comprised a number of 10-noded tetrahedral elements. The mesh global coarseness was first set to "Medium", then the local refinement factor was decreased to 0.8.

Calculation Phases:

- o **Initial Phase:** The Initial Phase represents the soil layers with no influence of concrete pad.
- o **Phase 1**: In this phase concrete pad with interface are introduced.
- o **Phase 2**: In this phase the loads are activated.

- **Step 1:** Firstly, the numerical model was calibrated with considerations of vertical load only and verified based on analytical estimates. The maximum displacement of the concrete pad based on numerical analyses was in the order of 0.1394m (139.4mm). This had a reasonable agreement with the analytical estimation of 156.8mm.

- **Step-2:** GRE (16 Numbers, 12m long, 4x4 grid – Only Vertical Load Case). The GRE's that were driven in soil to enhance capacity are concrete spun piles (0.4m diameter and 12 meters long). These along with the concrete pad were modelled as volume elements in Plaxis 3D with concrete properties.
    - **Initial Phase:** The Initial Phase represents the soil layers with no influence of GREs.
    - **Phase 1**: In this phase GREs are introduced as volume elements in soil layers.
    - **Phase 2:** In this phase concrete pad with interface are introduced.
    - **Phase 3**: In this phase, only vertical load is activated.

Plaxis 3D simulation:

The calibrated finite element model in PLAXIS 3D was subjected to **only vertical load of 639.68 kN** and the analysis was carried out with 16 numbers of GRE of 12 m long. The maximum displacement of the concrete pad is in the order of **0.01151m (11.51mm). Figure 3.4.3-2** shows the displacement contour of intake structure foundation with GRE for only vertical load condition.

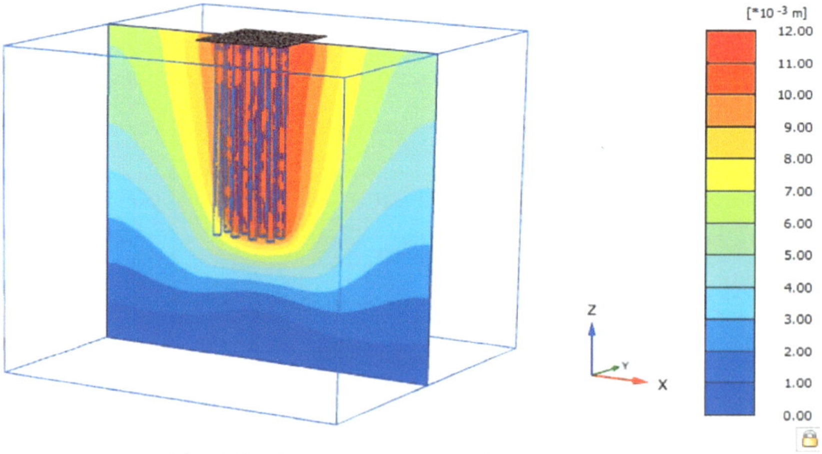

Total displacemnet [U] (scaled up 50.0 times)
Maximum value = 0.01151 m

Figure 3.4.3-2: Displacement contour of intake structure foundation with GRE (Vertical Load)

- **Step-3:** Without GRE (Vertical, Horizontal and Moment-VHM) Load Case)
  In this case, the calibrated finite element model was subjected to VHM loading as per the requirement. This analysis was carried out to check if there are any additional displacements due to the combination of loads. The analysis resulted in the vertical displacement of **0.3015m (301.5mm)**.
- **Step-4**: With GRE (16 Numbers, 12m long, 4x4 grid - VHM Load Case)

Plaxis 3D simulation

The Plaxis 3D model with 16 numbers of GRE and 12 m long were subjected to VHM loads specified earlier. The analysis resulted in a vertical displacement Uz of **19.50 mm**. **Figure 3.4.3-3** shows vertical displacement contour for VHM loads on foundation with 16 numbers of GREs,12m long.

- **Step-4A: With Optimized GRE (16 Numbers, 8m long, 4x4 grid - VHM Load Case).** The same analysis was performed by optimizing the length of the GREs to 8m. This analysis resulted in a vertical displacement of U_z of **31.41 mm**.

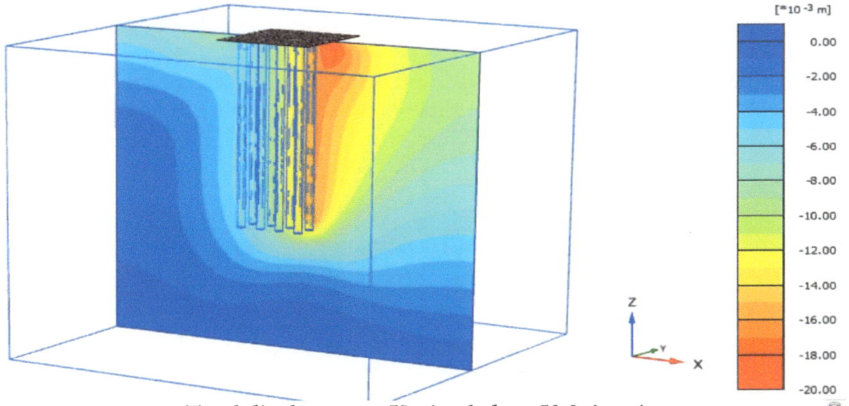

Total displacemnet Uz (scaled up 50.0 times)
Maximum value = 0.000 m
Minimum value = -0.01950 m

Figure 3.4.3-3: Displacement contour of the improved ground with 12m long GREs (V-H-M Load)

- **Step-4B: With Optimized GRE (9 Numbers, 12m long, 3x3 grid - VHM Load Case).** Another optimized design of the GREs was considered by reducing the number of GREs and maintaining the length as 12m. This analysis resulted in a vertical displacement U_z of **42.43 mm.**

Summary of Analyses

Table 3.4.3-3 presents the results of all the analyses performed to estimate the foundation settlements for the intake structure.

<table>
<tr><th>Ground condition</th><th>GRE Pattern</th><th>Settlement, mm</th><th>Remarks</th></tr>
<tr><td colspan="4">Vertical loading</td></tr>
<tr><td>Without GRE</td><td>-</td><td>139.4(156)*</td><td>-</td></tr>
<tr><td>With GRE</td><td>400mm OD, 12m long, 16Nos.</td><td>11.51</td><td>-</td></tr>
<tr><td colspan="4">VHM loading</td></tr>
<tr><td>Without GRE</td><td>-</td><td>301.5</td><td>-</td></tr>
<tr><td rowspan="3">With GRE</td><td>400mm OD, 12m long, 16Nos</td><td>19.50</td><td>Recommended</td></tr>
<tr><td>400mm OD, 8m long, 16Nos</td><td>31.41</td><td>-</td></tr>
<tr><td>400mm OD, 12m long, 9Nos</td><td>42.43</td><td>-</td></tr>
<tr><td colspan="4">*Analytical Estimate</td></tr>
</table>

Table 3.4.3-3: Summary of numerical analyses results

Pile Drivability Analyses

As per the recommendation from plaxis analysis, it was proposed to drive sixteen numbers 400mm OD spun piles to a final penetration of 11.5m for peripheral piles and 11.7m for internal piles using a steel follower and Junttan HHK7A hydraulic hammer. GRLWEAP 2014 Offshore version of the software was used to model the hammer-pile-soil system to evaluate the drivability in terms of dynamic pile stress and blowcounts vs penetration. The geometry considered for pile drivability analysis is shown in **Table 3.4.3-4.**

Pile	Outer dia (m)	Length (m)	Wall thickness (mm)	Target penetration (m)	Yield strength (MPa)	Modulus of elasticity (MPa)
Follower	0.356	8.00	11.13	-	240	210000
Spun Pile	0.400	12.0	75.00	11.70m (Below - 9.3EL.)	50	40000

Table 3.4.3-4: Geometry of Spun pile and Follower

A new 20mm thick plywood cushion having a modulus of elasticity of 207MPa was considered in Pile drivability analysis. The soil design parameters as per the geotechnical report as shown in **Table 3.4.3-5** were considered for estimating the soil resistance to driving (SRD).

Sl. No	Depth from EL-9.3m	Soil Type	γ^I (kN/m^3)	Shear Strength (kPa)	
				Top	Bottom
1	0.00-11.70	Clay	8.0	88	88

Table 3.4.3-5: Soil design parameters for estimating the SRD

GRLWEAP Analysis was performed by using the proposed JUNNTAN HHK7A on Spun pile (400mm diameter). The details of the hammer are summarized in **Table 3.4.3-6.**

Hammer Type	Hammer Model	Eq. Max. Stroke, m	Ram Weight, kN	Rated Energy, kN-m	Maximum Hammer Efficiency (%)
Hydraulic	Junntan-HHK7A	1.20	68.70	82.404	80

Table 3.4.3-6: Junntan-HHK7A hammer details

GRLWEAP offshore version software was used to model the hammer-pile-soil system. The proposed hammer, pile and the soils static and dynamic parameters were modeled to simulate the after effects of ram impact. The program computes the following:

1. The blow count (number of hammer blows/unit length of permanent set) of a pile under one or more assumed ultimate resistance values and other dynamic soil resistance parameters, given a hammer and driving system.

2. The axial stresses in a pile averaged over its cross-section associated with the assumed capacity value(s).
3. The energy transferred by the hammer to the pile for each capacity analyzed.

The soil resistance to driving was estimated for the following conditions,

a.) Continuous driving Lower Bound: It is estimated by considering 30% of static shaft capacity and the end bearing is assumed to act on the gross area.

b.) Continuous driving Upper Bound: It is estimated by considering 50% of static shaft capacity and the end bearing is assumed to act on the gross area.

c.) Full setup plugged condition: It is estimated by considering 100% of static shaft capacity and the end bearing is assumed to act on the gross area.

The soil dynamic parameters used to estimate the soil dynamic resistance are summarized in **Table 3.4.3-7.**

Soil Type	**Quake, mm**		**Damping, m/sec**	
	Side	**Toe**	**Side**	**Toe**
Clay	2.54	2.54	0.656	0.495

Table 3.4.3-7: Soil dynamic parameters considered for GRL-WEAP

Self-weight penetration was estimated by equating the weight of pile and weight of pile +weight of hammer to the soil resistance to penetration. The corresponding details are provided in **Table 3.4.3-8**. The Lower and Upper bound self-weight penetrations are estimated as below.

- Lower bound is estimated by considering 100% of static shaft capacity for clay layer whereas end bearing is assumed to act on the gross area.
- Upper bound is estimated by considering 50% of static shaft capacity for clay layer, and the end bearing is assumed to act on the gross area.

Spun Pile, kN	**Follower, kN**	**HHK7A, kN**
14.91	8.03	110

Table 3.4.3-8: Self Weight of the Spun pile, Follower and Hammer

The self-weight penetration results are summarized below in **Table 3.4.3-9.**

Criteria	Pile self-weight penetration (Vertical), m
	Spun pile + Follower+ HHK7A
LB	1.40
UB	2.30

Table 3.4.3-9: Summary of pile self-weight penetrations

Allowable Driving Stresses as per GRLWEAP

- The maximum allowable compressive stress for pre-stressed concrete is 0.85f'c-fpe
- The maximum allowable Tensile stress for pre-stressed concrete is fpe +0.25(f'c)1/2
- The maximum allowable compressive stress for pre-stressed concrete is 0.9fy

Where,

f'_c = concrete strength

f_{pe} = Effective pre-stress

f_y = yield strength of Pile material

The following details are used to estimate allowable compressive and tensile stress which are presented in **Table 3.4.3-10.**

Concrete strength = 50 MPa

Effective pre-stress = 4.18Mpa

Yield strength = 240MPa

Pile Material	Driving stress		Allowable Stress		Result
	Compressive (MPa)	Tensile (MPa)	Compressive (MPa)	Tensile (MPa)	
Concrete	28.2	5.3	38.32	5.96	**ok**
Steel	185.5	20.7	216	216	**ok**

Table 3.4.3-10: Summary of driving stress and allowable stress

Results of Drivability Analysis using GRLWEAP

- Junttan-HHK7A could drive the piles up to a target penetration of 11.7m with limiting hammer efficiency of 30%, 40%, 50% and 60% up to a depth of 3.0m, 5.0m, 9.0m and 11.70m respectively, with maximum blow counts of 4bl/0.25m, 5bl/0.25m and 8 bl/0.25m under continuous driving lower bound, continuous driving upper bound and Full set up conditions respectively.

- The maximum compressive and tensile stress expected during driving are 28.2MPa and 5.3MPa, respectively.

Conclusions

Settlement Assessment:

a) The intake head concrete base placed on the unimproved ground at EL -9.3m LAT was estimated to undergo a long-term settlement of 320.7mm. Infact, it had undergone failure upon placement of footing on the seabed. The proposed ground improvement with GREs comprising 16 no's of 12m long 400mm OD spun piles resulted in a reduced vertical settlement of **19.50 mm** which was adequate meeting the client's requirement and thus the same was recommended.

b) The additional settlement analyses considering the ground improvement with the reduced number and length of GREs resulted in more than 40mm settlement.

Pile Drivability:

The Spun Pile plus Follower plus Hammer was expected to self-penetrate to a depth ranging of 1.40 – 2.30 m. Hammer efficiency was recommended to be adjusted in-line with pile penetration during initial driving. Junttan HHK7A was predicted to drive the spun pile to the intended final penetration of 11.70m with limiting hammer efficiency as shown in **Table 3.4.3-11.**

Depth, m	Limiting Hammer Efficiency, %
0.00 - 1.40	Self-weight Penetration
1.40 - 3.00	30
3.00 - 5.00	40
5.00 - 9.00	50
9.00 - 11.70	60

Table 3.4.3-11: Limiting Hammer Efficiency

The maximum compressive and tensile stresses expected while driving the concrete spun pile with the above limiting efficiencies were 28.2 MPa and 5.3MPa and these stresses were within the allowable limits.

The GREs were driven by Junttan hammer, and the intake structure was successfully placed on the concrete pad over the improved ground.

3.5 Associated Risks and Engineering Solutions

3.5.1 Mud mats

In the South-East Asian region and particularly in offshore Indonesia, the mudline is often characterized by soft cohesive materials with shear strengths well below 5 kPa.

Retrieving accurate samples of such materials poses a significant challenge due to their delicate nature. Failure to secure proper samples can result in inaccuracies in determining shear strength parameters for mudmat designs. Such inaccuracies could result in catastrophic events like Mudmat failure and Jacket tilting.

To mitigate these risks, it is imperative to employ specialized samplers designed for retrieving high-quality samples from soft mudline sediments. Attempts should be made to obtain high quality samples by appropriate sampling techniques, i.e., liner samplers.

On-board geotechnical laboratories play a pivotal role in this process, where samples are tested using methods such as Torvane and Miniature vane testing to ascertain their shear strength characteristics. These testings shall be performed by trained and experienced personnel. Further, the shear strength can also be estimated by performing in-situ tests such as CPT, T-Bar, and Ball Penetrometers.

The data obtained from these tests forms the basis for determining the optimal design profile for mud mats supporting offshore jackets. Design engineers rely on this information to develop robust designs including skirts that can withstand the challenging conditions posed by soft mudline sediments.

Numerical analyses are often conducted to predict mudmat settlements, tilt, and stability. In essence, the accurate characterization of shear strength parameters of soft mudline sediments is paramount for ensuring the structural integrity and stability of offshore installations.

Through meticulous sampling, testing, and analysis, engineers can devise resilient designs capable of withstanding the dynamic forces present in South-East Asian marine environments.

3.5.2 Pile Foundations

The geological composition typically consists of uniform layers of cohesive soil interspersed with thin sand layers necessitating meticulous planning to ensure the safe and effective installation of piles. When it comes to pile driving operations, the use of a standard hydraulic hammer generating 200 to 500 kJ of energy is deemed sufficient to drive pile sections to depths ranging from 120 to 130 meters below the mudline. Despite occasional encounters with sand layers at various depths, these layers do not present significant resistance to pile driving, but an increase in blow counts is observed. However, it is necessary to ensure that the piles are installed to sufficient depths to support the intended load-carrying capacity. Pile driving in cohesive materials will remold the soil resulting in reduced capacity at the end of drive vis-à-vis static resistance. Pile monitoring by instrumenting the pile with strain gauges and accelerometers, will aid in estimating the available capacity at the end of driving. A restrike test will estimate the increase in capacity with the time which is a setup phenomenon. Using the re-strike setup, long-term capacity can be estimated using the available empirical relationship of setup and time e.g., Bullock et al (2005).

3.5.3 Leg Penetration Analysis for Jackup Rigs

When conducting site-specific assessments for the positioning of Jackup rigs in offshore Indonesia, an essential aspect is the leg penetration analysis. This assessment is performed to estimate the Jackup leg penetration and potential hazard due to punch-through. The punch-through hazard occurs when the spud can penetrate through a soil layer with higher shear strength (desiccated layer) and into an underlying layer with lesser shear strength. Such punch-through incidents can compromise the rig's stability and lead to catastrophic accidents. Therefore, it is imperative to evaluate the shear strength profile in both crustal and underlying weak layers by conducting a thorough geotechnical investigation. Information from geophysical surveys will further aid in mitigating hazards such as channel features which are not uncommon in offshore environments.

Installation of Pipelines

In general, the seabed soil is very soft to soft in consistency. However, firm to stiff cohesive soils are not uncommon at certain chainage along the pipeline route.

Zoning the soil profiles along the pipeline will assist in evaluating the burial or otherwise. Free span correction and/or pipeline supports can also be planned.

Liquefaction

Liquefaction may occur in cohesionless soils. In this region, the soil layers consist of very soft to soft clays. These clays are generally not susceptible to liquefaction, except some near shore locations where sandy seabed exists.

Scour

Loose cohesionless seabed sands are susceptible to scour due to high seabed currents. In general, the offshore Indonesia seabed soils consist of cohesive soils, and thus scour effects are unlikely.

3.6 Lessons from the Experienced

3.6.1 From North Sea to Southeast Asia: Leadership Insights from Gary Selbie, Harbour Energy's President Director.

Synopsis

The interview provides valuable insights into the evolution of oil and gas operations in Southeast Asia, emphasizing the importance of local knowledge, continuous learning, and adaptation to change.

Gary's journey from a drilling engineer to a senior leader demonstrates the value of embracing challenges, learning from mistakes, and maintaining a balance between professional excellence and personal well-being.

Introduction

In this insightful interview, a seasoned oil and gas industry leader shares his 30+ year journey from a drilling engineer to a senior leadership position.

The conversation reveals valuable lessons about adapting to different geographical challenges, the importance of local expertise, and the evolution of safety practices in geotechnical investigations across Southeast Asia.

Interviewee Profile

Gary Selbie is a senior leader at Harbour Energy (formerly Premier Oil), with over three decades of experience in the oil and gas industry. Beginning his career in 1990 as a trainee drilling engineer, he joined Premier Oil in 1997 and has since played a pivotal role in transforming Indonesia's Natuna Sea Block A from a small oil facility into a multi-asset gas facility. His expertise spans drilling operations, project management, and leadership across various geographical regions, with particular emphasis on Southeast Asian operations.

The Talk

CRP: Dr CR Parthasarathy

GS: Gary Selbie, President Director, Harbour Energy, Jakarta, Indonesia

CRP: Could you share your career journey leading to your current leadership position?

GS: I started my career in 1990 as a trainee drilling engineer with a service provider. After seven years with contractors, I joined Premier Oil in 1997 and have remained with the company through its transition to Harbour Energy, progressing through roles including drilling superintendent, drilling manager, and corporate drilling manager.

CRP: How has the transition from Premier Oil to Harbour Energy been?

GS: While the change from being a relatively small independent to part of a much larger company has required adjustment, it has brought benefits such as improved access to funding and the ability to pursue larger-scale projects and developments. The transition was a natural progression for Premier, opening new opportunities.

CRP: What significant projects have shaped your career?

GS: Indonesia has been a major part of my career. Since 1997, I've probably spent 20 years in this region. The Natuna Sea Block A field has been incredibly significant for me. When we took over in 1997, we recognized the potential to transform it from an oil field to a gas field, and we installed a gas pipeline.

I was personally involved in all the key developments - the infill drilling on Anoa, West Lobe, Natuna sea, and Gajah Baru. I've seen the Block A field grow from a fairly small oil facility to a multi-asset gas facility over the last 25 years. It's almost like my baby. I'm proud that it's become the backbone of Premier's - and now Harbour Energy's - long-term business in Indonesia.

CRP: What role does geotechnical investigation play in oil and gas construction projects?

GS: Geotechnical investigations and pre-planning are critical, particularly for Jack up operations. We learned valuable lessons after experiencing a major punch-through incident with the Harvey Ward. The industry has since refined its approach, requiring specific rig locations and boreholes on each spud can. This learning curve has benefited both Indonesian and Malaysian operations.

CRP: What challenges have you faced in your leadership journey?

GS: As a driller, I've seen my fair share of site surveys, geophysics, and geotech work, and yeah, we've had our moments. Speaking of incidents, there's the Harvey Ward. Back then, and I'm talking about the Natuna Sea area, we just didn't fully grasp the punch-through risk. That was a real eye-opener.

Then, more recently, in Myanmar, we were trying to get a jack-up next to an existing platform. Even with the best industry experts on board, doing the surveys, setting the preload sequence, the whole nine yards, we hit a wall. The spud cans just wouldn't punch through this hard crust. We were stuck, couldn't increase the preload, and were looking at a potential catastrophic punch-through. Weeks went by, a real stalemate, trying to figure out how to get that jack-up in place and finish the job. Honestly, that one was a real kick in the teeth. We thought we had it covered, experts and all, but we still ended up with no solution. It just goes to show, you know, you learn from every job. I've taken away a ton from those experiences, both from the operational side, and from seeing how even the best geotechnical guys can get caught out.

It really underlines how critical thorough pre-planning and understanding the ground conditions are, especially with jack-ups. You can't afford to take anything for granted.

CRP: What advice would you give to upcoming industry professionals?

GS: Local knowledge is crucial. Coming from the North Sea, which was considered the pinnacle of oil and gas technology, I learned that solutions that work in one region may not work in another. It's essential to listen carefully to local experts and keep an open mind when working in different geographical contexts.

Additionally, I recognize the role of mentorship in professional development. I have had the opportunity to learn from several individuals who have provided guidance. Mentorship has been key to my development. Tim Hanson, my drilling monitor, taught me technical skills and people management. I value his guidance.

CRP: What is the industry outlook for next 12 months and 5 years in the region?

GS: We are observing a marked increase in activity across the region. Rig and vessel utilization, alongside personnel engagement, have all seen rapid growth over the past year and a half. This heightened activity level is expected to continue. While the energy transition is unavoidable, the immediate future indicates a substantial surge in gas exploration and development, particularly given the recent energy challenges. We anticipate a very active year ahead, with follow-on appraisal work in the deep-water area of the Andaman Sea. If the exploration and appraisal wells proceed as planned, further appraisal and development will follow. Consequently, Indonesia and the broader region are positioned for sustained activity over the next five years, at least.

CRP: How do you maintain work-life balance in your leadership role?

GS: Since the pandemic, we've adopted a hybrid working model with three days in the office and two days from home. I typically start around 8:00-8:30 AM and work later to accommodate London office hours.

I maintain physical health through weekly golf and swimming, which the hybrid working model has made easier to schedule.

CRP: Could you tell us about your family life?

GS: I met my Indonesian wife after arriving here in 1997, and we've been married for over 20 years. We have one daughter, aged 19 and a son aged 16, who receive their education in the U.K. Managing an international career while maintaining family connections has been challenging, especially during the pandemic, but we've managed to balance professional and personal aspects effectively.

3.6.2 Drilling Deep: Syukri Fitrialdi's Vision for Geotechnical Excellence in Indonesia

Synopsis

Syukri Fitrialdi's journey with Taka Hydrocore Indonesia illustrates the critical role of specialized technical services in driving industrial progress. His leadership demonstrates how technical expertise, strategic vision, and adaptability can transform challenges into opportunities.

Introduction

In an era of complex industrial challenges, leaders who combine technical expertise with strategic vision stand out. Dr. CR Parthasarathy's exclusive interview with PAK Syukri Fitrialdi reveals the remarkable journey of a pioneering entrepreneur in Indonesia's geotechnical services sector.

Interviewee Profile

Syukri Fitrialdi is the President Director of Taka Hydrocore Indonesia (THI), a pioneering geotechnical services company he founded in 2010.

With over two decades of experience in the technical services sector, Fitrialdi has established himself as a key player in Indonesia's industrial landscape.

The Talk

CRP: Dr CR Parthasarathy

SF: Syukri Fitrialdi, President Director, Taka Hydrocore Indonesia.

CRP: Can you please share with us the important events or experience or the projects that have shaped your life and career so far?

SF: I initiated my business journey from ground zero. One significant project I spearheaded from inception was the development of a major LNG facility in Indonesia back in 1994. From its inception, I was deeply involved in every aspect, continuously refining inputs and processes throughout its lifecycle. Today, I am proud to say that the project has reached fruition, successfully delivering LNG to the Malaysian market and beyond. This experience, starting from scratch and witnessing the project through to completion, has been invaluable and deeply fulfilling.

CRP: Being in this field, how relevant do you think is the geotechnical investigation for the success of any oil and gas projects?

SF: Geotechnical analysis serves as the foundation for risk mitigation in all oil and gas structures. It's mandatory for every structure to undergo thorough and dependable geotechnical assessment. Consequently, the industry relies heavily on accurate data to ensure the integrity and safety of these structures.

CRP: What are your philosophy principles and what processes do you adopt in the company?

SF: In Taka, we uphold the 'SIGAP' philosophy, emphasizing SERVICE (S), Integrity (I), Growth (G), Awareness (A), and People Development (P) as integral elements of our company culture.

This philosophy is consistently evolving and integrated into our systems, shaping the fabric of our organization.

CRP: Were there any challenges you have faced in this long journey and what are the lessons that you have learned from?

SF: Yes, we faced a challenging period during the low oil prices around 2016-17, nearing bankruptcy due to a lack of projects despite prior substantial investments. To overcome this adversity, we strategically began by undertaking smaller projects, gradually rebuilding and ultimately achieving significant growth. This particular year marked a difficult phase for us amid the industry's lowest prices.

CRP: do you have any tips or lessons for the readers of the book?

SF: Don't lose hope. When faced with challenges, consider them as opportunities for growth. Stay resilient and focus on achieving your goals.

CRP: Fantastic. Thank you for sharing that. What is the industry outlook for the next 12 months? And how do you see oil and gas in the next five years?

SF: I believe that the Oil and Gas industry will continue to expand over the next five years. Despite the emergence of green energy, the demand for oil and gas remains robust. While the transition to a more balanced energy landscape is ongoing, it is not expected to be fully achieved within the next five years. Therefore, there are still opportunities for individuals to enter the oil and gas industry.

CRP: What's your vision and plan for the future?

SF: We must adapt to regional conditions, particularly in response to changing demands and business landscapes. For instance, I anticipate that in five years, green energy will dominate, and we must align our strategies accordingly. We are ready to embrace this shift.

3.6.3 Leading Through Innovation: A Conversation with Mr Vincent Chan on Meindo's Journey in Oil & Gas Construction

Synopsis

Vincent Chan's leadership journey at MEINDO exemplifies the successful transition of a family business into a modern, forward-thinking EPCI company. His emphasis on methodical growth, technical excellence,

and adaptation to challenges positions MEINDO well for its global ambitions.

The interview highlights the critical balance between maintaining traditional strengths while embracing innovation and sustainability in the evolving oil and gas sector.

Introduction

In an insightful interview, Vincent Chan, second-generation leader of MEINDO, shares the company's remarkable evolution from a civil contractor to a leading EPCI (Engineering, Procurement, Construction, and Installation) player in Indonesia's oil and gas sector. The discussion reveals crucial insights into geotechnical investigation's role in EPC projects, leadership challenges during COVID-19, and the company's vision for global expansion.

Interviewee Profile

Vincent Chan serves as a CEO at MEINDO, a prominent Indonesian EPCI company founded in 1987. Following in his father PAK Chan Jemmi's footsteps, Vincent joined the company after completing his engineering education. Under his leadership, MEINDO has expanded its capabilities from civil contracting to comprehensive oil and gas services, with significant investments in manpower and marine capabilities.

His engineering background and hands-on experience, starting as a project engineer, have shaped his approach to leading the company through challenging times, including the COVID-19 pandemic.

The Talk

CRP: Dr CR Parthasarathy

VC: Pak Vincent Chan, Operations Director, Pt Meindo Elang Indah, Jakarta, Indonesia

CRP: Could you tell us about MEINDO's journey so far?

VC: MEINDO was founded in 1987 as a civil contractor. The company expanded into oil and gas in the early 2000s, gradually building engineering capabilities and earning client trust. Through conservative, step-by-step growth and significant investments in manpower and marine capabilities, MEINDO has achieved its current scale.

CRP: How did you build your career to reach this leadership level?

VC: My career path to leadership started with a mix of family legacy and personal choice. My father founded this business, and after earning my engineering degree, he encouraged me to join him. Initially, I hesitated—I wanted to build my own career—but I trusted his vision and decided to take the leap. Starting as a project engineer, I learned the ropes by diving into technical details, understanding how engineering disciplines intersect with business goals. Early on, I realized the importance of non-negotiable standards: safety, quality, and sustainability aren't just checkboxes—they're the foundation of trust in our industry. Collaborating with partners like Sarathy reinforced how critical teamwork and shared values are to delivering results.

My father has been my biggest influence. He taught me that hard work and critical thinking solve most problems. In projects, challenges are inevitable—no job is problem-free but approaching them methodically always leads to solutions. I've also learned from mentors and peers along the way, picking up habits like disciplined note-taking and clear communication, which keep our teams aligned and efficient.

Today, I lead with the same principles: focus on excellence, embrace challenges as opportunities, and never cut corners. It's about balancing legacy with innovation, and ensuring our team grows while staying true to the standards that got us here.

CRP: What were the most significant challenges or defining moments in your career journey, and what lessons did you learn from them that shaped your growth into a leadership role?

VC: Early in my career, a project was under time pressure. This led to overlooked engineering details and inefficiencies —a lesson in prioritizing preparation over speed. The lesson was clear: cutting corners on planning, even under time constraints, compromises outcomes. It reinforced the value of disciplined preparation, no matter the urgency.

The COVID-19 pandemic proved particularly challenging. Managing projects during this period of unprecedented uncertainty taught valuable lessons about business planning for unknown scenarios. This experience has better prepared the company for future uncertainties.

CRP: How do you rate the importance of geotechnical investigation for EPC projects?

VC: Geotechnical investigation is crucial for EPCI projects, particularly in offshore operations. Poor investigations can lead to significant challenges, such as encountering harder soil conditions during pipe laying and pile driving operations. Quality geotechnical data collection, laboratory analysis, and design work are fundamental to project success.

CRP: What principles and practices keep MEINDO ahead in the industry?

VC: The company follows the motto 'committed and reliable,' focusing on completing projects while maintaining quality, safety, and regulatory compliance. MEINDO benchmarks against global competitors and emphasizes digitalization to improve productivity and optimize solutions.

CRP: What is your perspective on the industry's future?

VC: Despite environmental concerns, I see a positive outlook for oil and gas, particularly in Indonesia, with targets of one million barrels per day and 12 million cubic feet per day of production by 2030. I emphasize the importance of developing greener operations while maintaining stable energy production.

CRP: What are MEINDO's future plans?

VC: The company aims to become a global EPCI player, having already completed international projects in Vietnam during COVID-19. They continue to compete for international projects while expanding their global reach.

CRP: How do you balance the demands of your leadership role with personal interests and family life?

VC: On workdays, I typically start early (in the office by 8 AM) and often stay late due to the demands of leadership, which spans finance, operations, and strategic planning. Despite this, I prioritize returning home to spend time with my family, balancing weekends for compensation. My hobbies include soccer, futsal, golf, reading, and connecting with friends, clients, and partners.

Family plays a central role; most of us are engineers (my sister is an accountant), and we're deeply close-knit. Work naturally dominates conversations, even during weekends or gatherings, which keeps us aligned on business goals. These dynamic fosters collaboration and quick decision-making, ensuring we're collectively focused on the company's vision.

4
MALAYSIA

4.1 Introduction

The beginning of Malaysia's petroleum industry can be traced to early in the century, when oil was first discovered in Miri, Sarawak. The Malaysian offshore area **(Figure 4.1-1)** covers about 364,450 km2, of which 264,950 km2 consists of shallow continental shelf (water depth < 200 m) and the remaining 99,500 km2 (or 27%) is made up of deep-water areas with depths of up to more than 2000m.

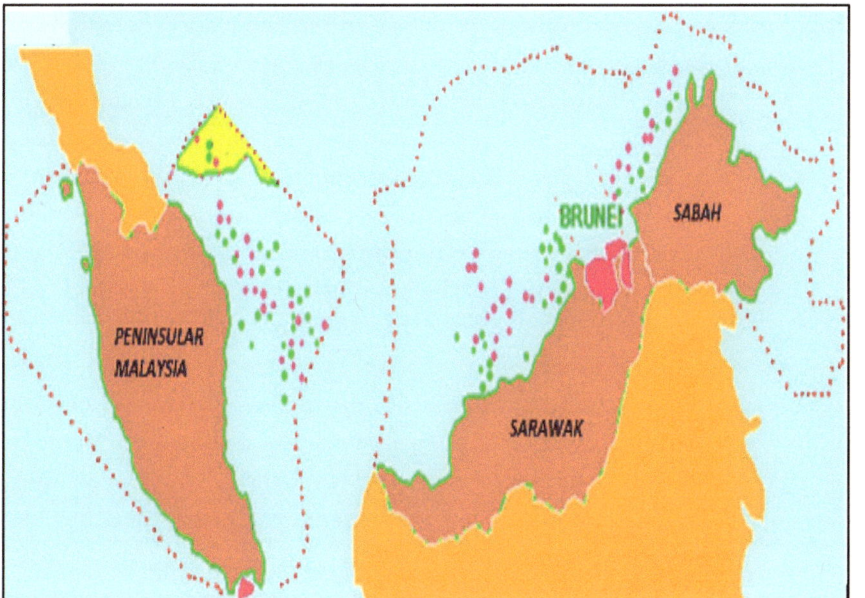

Figure 4.1-1: Offshore Regions in Malaysia (Bashir et al. 2018)

The continental shelf is underlain by thick Tertiary sedimentary sequences and occupies a significant portion of the Sunda Shelf, one of the largest continuous shallow water areas in the world. Exploration activities offshore of Malaysia have been concentrated mainly in the three major sedimentary basins - the Malay Basin, the Sarawak Basin, and the Sabah Basin.

4.2 Geology of Offshore Regions in Malaysia

4.2.1 Tectonic Setting

The offshore regions of Malaysia are part of the Sunda Shelf, a shallow continental shelf extending from the Malay Peninsula to Borneo. These regions are influenced by the interactions of several tectonic plates, including the Eurasian Plate, Indo-Australian Plate, and Philippine Sea Plate. The collision and subduction of these plates, such as along the Sunda Trench off Sumatra, have shaped the region's geology. During the Cenozoic era, extensional tectonics led to the formation of sedimentary basins, which are now the primary focus of oil and gas exploration.

4.2.2 Major Sedimentary Basins

Malaysia's offshore regions are characterized by several sedimentary basins formed through rifting, subsidence, and sedimentation over millions of years. These basins are rich in hydrocarbons and include:

- **Malay Basin**: Located in offshore Peninsular Malaysia in the Gulf of Thailand and South China Sea, this basin was formed during the Oligocene to Miocene periods. It contains thick sequences of fluvial, lacustrine, and marine sediments, with major hydrocarbon-bearing formations such as Group K and Group L sandstones. The basin is known for its gas fields and some oil reserves (Riehle et al. 1999).
- **Sarawak Basin**: Situated offshore Sarawak in the South China Sea, this basin was formed during the Late Eocene to Miocene. It comprises deltaic and shallow marine sediments, with major hydrocarbon reservoirs in the Cycle I and Cycle II formations. The basin is renowned for its large gas fields, including the Bintulu LNG projects.
- **Sabah Basin**: Located offshore Sabah in the South China Sea and Sulu Sea, this basin was formed during the Late Oligocene to Miocene. It contains deepwater turbidite systems and

carbonate buildups, with major hydrocarbon reservoirs in the Sandakan and Kudat formations. The basin is known for its oil and gas fields, including deepwater discoveries.

- **Penyu Basin**: Found offshore Peninsular Malaysia, east of the Malay Basin, this basin was formed during the Late Cretaceous to Paleogene. It contains fluvial and lacustrine sediments and, although less explored, has potential for hydrocarbons.
- **Natuna Basin**: Located offshore Sarawak near the Indonesia-Malaysia border in the South China Sea, this basin was formed during the Eocene to Miocene. It contains carbonate reefs and clastic sediments and is known for its large gas fields, including the East Natuna Field.

4.2.3 Structural Features

The offshore regions of Malaysia are marked by several structural features that influence hydrocarbon accumulation. Extensional faults and rifts are common in the basins, creating traps for hydrocarbons. Carbonate platforms, especially in Sabah and Sarawak, are extensive and serve as reservoirs.

Deepwater areas, such as those offshore Sabah, are characterized by turbidite systems and submarine fans, which are increasingly targeted for exploration.

4.2.4 Hydrocarbon Systems

Malaysia's offshore basins are rich in hydrocarbons due to well-developed source rocks, reservoirs, seals, and traps. Source rocks include organic-rich shales and coals deposited in lacustrine and deltaic environments, such as the Group K shales in the Malay Basin. Reservoirs are primarily sandstones (e.g., fluvial and shallow marine sands) and carbonates (e.g., reefs and platforms). Seals consist of shales and mudstones that cap the reservoirs, while traps are both structural (e.g., fault blocks, anticlines) and stratigraphic (e.g., pinch-outs, reefs).

4.2.5 Geological Hazards

The offshore regions of Malaysia are not without geological hazards. Earthquakes are a concern, particularly near Sabah, due to the collision of the Eurasian and Philippine Sea Plates. Subduction-related volcanism affects regions near the Sunda Trench, and there is a potential risk of tsunamis from undersea earthquakes in the Andaman Sea and South China Sea.

4.2.6 Deepwater Exploration

Deepwater exploration is a growing focus in Malaysia's offshore regions, particularly in Sabah and Sarawak. These areas feature submarine canyons, turbidite systems, and carbonate mounds. Major deepwater oil and gas fields, such as Gumusut-Kakap and Kikeh, have been discovered, highlighting the potential of these regions (Riehle et al. 1999).

4.2.7 Environmental and Economic Significance

The offshore regions of Malaysia are of immense environmental and economic importance. They are among the most prolific oil and gas producers in Southeast Asia, contributing significantly to the country's economy. Additionally, these regions are home to rich biodiversity, including coral reefs and marine ecosystems, particularly in areas like the Spratly Islands and Sulu Sea. However, overlapping territorial claims in the South China Sea pose challenges for exploration and resource development.

The offshore geology of Malaysia is dominated by sedimentary basins formed through tectonic rifting and subsidence. These basins are rich in hydrocarbons, making Malaysia a major oil and gas producer in the region.

The interplay of tectonic forces, sedimentation, and sea-level changes has created diverse geological features, from shallow carbonate platforms to deep-water turbidite systems. Understanding this geology is crucial for resource exploration and environmental management. For a concise overview of Malaysia basins see below **Table 4.2.7-1**:

Aspect	Details
Tectonic Setting	Part of the Sunda Shelf; influenced by Eurasian, Indo-Australian, and Philippine Sea Plates.
Major Basins	Malay Basin, Sarawak Basin, Sabah Basin, Penyu Basin, Natuna Basin.
Structural Features	Faults, rifts, carbonate platforms, deep-water turbidite systems.
Hydrocarbon Systems	Source rocks (shales, coals), reservoirs (sandstones, carbonates), seals (shales), traps (structural, stratigraphic).
Geological Hazards	Earthquakes, subduction-related volcanism, tsunami risks.

Deepwater Exploration	Focus on Sabah and Sarawak; features include submarine canyons, turbidites, and carbonate mounds.
Economic Significance	Major oil and gas production; contributes significantly to Malaysia's economy.
Environmental Importance	Rich biodiversity, including coral reefs and marine ecosystems.
Geopolitical Issues	Overlapping territorial claims in the South China Sea.

Table 4.2.7-1 : Overview of Malaysia Basins

4.3 Geotechnical Characteristics of Malaysia's Offshore Regions

The geotechnical characteristics of Malaysia's offshore regions are critical for engineering projects such as oil and gas platform installations, pipeline routing, and subsea infrastructure development.

These characteristics are influenced by the region's geology, sediment types, and environmental conditions. Below is a detailed overview of the geotechnical properties of Malaysia's offshore regions:

4.3.1 Sediment Types

The offshore regions of Malaysia are characterized by a variety of sediment types, which influence their geotechnical behavior:

- **Clay and Silt**: Dominant in many offshore basins, especially in the Malay Basin and Sarawak Basin. These fine-grained sediments have low permeability and high compressibility.
- **Sand and Gravel**: Found in shallow marine and deltaic environments, particularly in the Malay Basin and Sabah Basin. These coarse-grained sediments have higher permeability and shear strength.
- **Carbonate Sediments**: Common in regions with coral reefs and carbonate platforms, such as offshore Sabah and Sarawak. These sediments can be highly variable in strength and stability.
- **Organic-rich Sediments**: Present in deltaic and lacustrine environments, particularly in the Malay Basin. These sediments are often soft and have high organic content, leading to low shear strength.

4.3.2 Soil Properties

The geotechnical properties of offshore soil vary depending on the sediment type and depositional environment:

The offshore regions of Malaysia present several geotechnical challenges for engineering projects:

a) **Soft Clay and Silt**: Soft clays and silts have low shear strength and high compressibility, making them prone to settlement and instability.

b) **Carbonate Soils**: Carbonate sediments can be highly variable, with potential for weak, porous, or cemented layers. They are also prone to dissolution and erosion.

c) **Shallow Gas and Gas Hydrates**: Shallow gas accumulations and gas hydrates can cause sudden soil instability and pose risks during drilling and construction.

The geotechnical characteristics of Malaysia's offshore regions are diverse and complex, influenced by sediment types, geological history, and environmental conditions. Key challenges include soft clays, variable carbonate soils, shallow gas, slope instability, and cyclic loading (Moh 1987). Addressing these challenges requires detailed site investigations, specialized engineering solutions, and careful consideration of environmental impacts. Understanding these geotechnical properties is essential for the safe and efficient development of Malaysia's offshore resources.

4.4. General Geology of the Region

Mesozoic rock formations are found widespread in the Northwest and Central Belt of Peninsular Malaysia as well as in West Sarawak. Very few Mesozoic rock formations are known in the Eastern Belt of Peninsular Malaysia and in Sabah. Thirty-eight Mesozoic rock formations are known in Malaysia, including 28 formations from onshore Peninsular Malaysia, 8 formations from West Sarawak and 1 formation each from Sabah (the Madai-Baturong limestone) and the offshore Malay Basin (the Sotong limestone), (Peng et al. 2004).

Deepwater fold-and-thrust belts form both active and passive margins and often host compressional anticlines that trap hydrocarbons (Hesse et.al 2010).

The Sabah fold and thrust belt is an established circum-Borneo play with multiple commercial oil and gas fields located in this proven prolific geological province. The thrusts are likely formed from the combination of two geological processes. Firstly, as the result of subduction of the Dangerous Grounds beneath continental Borneo. Secondly, due to large-scale gravity sliding where up to seven kilometers of Miocene and younger clastic sediments were rapidly deposited via river deltas to the Sabah shelf and shelf-edge area.

Offshore oil and gas fields in Malaysia can generally be grouped based on their location and operator. Here's a generalized breakdown:

4.4.1 Sabah Offshore Fields:

- Gumusut-Kakap Field: Located in deep waters offshore Sabah, Gumusut-Kakap is one of Malaysia's largest oilfields. Operated by Petronas, it employs advanced deepwater drilling and production technologies.
- Kikeh Field: Another major deepwater oilfield offshore Sabah, operated by Murphy Oil. Kikeh utilizes advanced subsea infrastructure for production from its deepwater reservoirs.
- Malikai Field: Situated offshore Sabah, Malikai is operated by Shell Malaysia. It involves the production of oil and gas from deepwater reserves using innovative drilling and production techniques.

4.4.2 Sarawak Offshore Fields:

- Baram Field: Offshore Sarawak, operated by Petronas Carigali. Baram has been a significant contributor to Malaysia's oil production for years, undergoing multiple development phases.
- Larak Gas Field: Also at offshore Sarawak, operated by Petronas Carigali. Larak primarily produces natural gas, supporting Malaysia's domestic gas supply and industrial needs.
- Block SK318 (including Kebabangan and Kamunsu East fields): This offshore block in Sarawak includes fields such as Kebabangan and Kamunsu East, operated by Petronas Carigali and Shell, respectively. These fields contribute to Sarawak's offshore hydrocarbon production, utilizing advanced drilling and production technologies.

4.4.3 Peninsula Malaysia Offshore Fields:

- Tapis Field: Offshore Peninsula Malaysia, operated by ExxonMobil. Tapis is one of Malaysia's oldest and largest oilfields, employing advanced technologies for efficient production.
- Other smaller fields offshore Peninsula Malaysia: Various smaller fields contribute to Malaysia's offshore production in this region, operated by different companies and contributing to the country's oil and gas output.

4.4.4 Other Offshore Blocks:

- Block PM302: Located in the South China Sea, Block PM302 includes several oil and gas fields operated by companies like Petronas Carigali and JX Nippon Oil & Gas Exploration. These fields contribute to Malaysia's hydrocarbon reserves and production capacity.
- Other blocks in the South China Sea: Besides PM302, there are other offshore blocks in the South China Sea undergoing exploration and production activities, contributing to Malaysia's offshore oil and gas sector.

4.4.5 Gas Fields:

- Berantai Field: Offshore Terengganu, operated by Petronas Carigali. Berantai primarily produces natural gas and condensate, supporting Malaysia's domestic energy needs and exports.

The Malay Basin is the largest province of the Peninsular Malaysia. The other provinces include straits of Melaka, Arong Graben, Penyu Basin and some parts of the onshore area. There are six major fields in the area such as Seligi, Jerneh, Duyong, Guntong etc. The basin has been successfully explored in the last 50 years with 13-billion-barrel equivalent of oil (Bboe) has been discovered (Madon 2021).

4.5 General Stratigraphies of Offshore Malayasia

The following section provides a basic understanding of the typical stratigraphy of some the most important offshore oil and gas fields in Malaysia. The typical physical and engineering parameters are also presented in the form of geotechnical stratigraphy table as shown in **Table 4.5.1-1 and Table 4.5.2-2.**

4.5.1 Sabah-Sarawak

Depth (m)	Description	Shear Strength, S_u (kPa)		Angle of Internal Friction (phi)
		(Top)	(Bottom)	
0 – 2	Very Soft Lean Sandy CLAY	2	10	-
2 – 20	Dense SAND	-	-	33
20 – 25	Soft CLAY	5	20	-
25 – 30	Dense SAND	-	-	33
30 – 32	Stiff CLAY	25	50	-
32 – 35	Medium Dense SAND	-	-	32
35 – 52	Firm to Stiff CLAY	40	75	-
52 – 55	Dense SAND	-	-	33
55 – 88	Very Stiff CLAY	50	125	-
88 – 98	Medium Dense SAND	-	-	32
98 – 100	Very Stiff CLAY	100	125	-
100–105	Medium Dense SAND	-	-	32
105–135	Stiff to Hard CLAY	75	150	-
135–139	Medium Dense SAND	-	-	32
139–150	Stiff to very stiff CLAY	75	125	-

Table 4.5.1-1: Generalized Soil Stratigraphy of Sabah-Sawawak

4.5.2 Kelantan-Terrangannu

Depth (m)	Description	Shear Strength, Su (kPa)		Angle of Internal Friction (phi)
		(Top)	(Bottom)	
0 – 2	Very Soft CLAY	2	10	-
2 – 19	Firm to Very Stiff CLAY with traces of cemented layers	20	175	-
19 – 23	Very stiff CLAY	100	200	-
23 – 40	Very Stiff CLAY	125	175	-
40 – 44	Hard CLAY	200	200	-
44 – 48	Stiff to Very Stiff CLAY	75	175	-
48 – 52	Very stiff to Hard CLAY	100	200	-
52 – 105	Very Stiff CLAY	175	175	-
105–150	Very stiff to hard CLAY	125	250	-

Table 4.5.2-2: Generalized Soil Stratigraphy of Kelantan-Terengganu

4.6 Case Studies

4.6.1 Case Study – 1: Offshore Underwater Pile Monitoring (Dr Ramesh Gangisetti, Mr Soumik Gosal, Mr kris Gustavian & Mr Nugi)

Problem Statement

Pile driving causes the soil around the pile to lose strength, particularly cohesive soils, due to increased pore pressure and remolding. Once pile driving is complete, the cohesive soils regain strength due to consolidation and the phenomenon is called setup. Usually, if the soil profile is predominantly cohesive, the available pile capacity at the end-of-driving can only be around 25 to 35% of long-term static capacity. This case study explains how pile monitoring data of end-of-driving and the re-strike test can be used to estimate the long-term setup using the analytical model.

Project Details

It was proposed to install a 4-leg jacket platform at an offshore field in Malaysia having a water depth of 143m, offshore Malaysia. The pipe piles were made of single pile sections of 98.35m long and 2.438m diameter to be driven underwater to a final penetration of 80m. An IHC S-800 hydraulic hammer was proposed to drive piles.

The site-specific geotechnical borehole indicates a 6m thick, soft clay at the seabed followed by soft to firm clay up to 9m. Then firm to stiff clays follow up to 30m underlain by stiff to very stiff clays up to the depth of interest of 80m. The self-penetration of piles ranged from 23 to 26m. Pile monitoring using the Pile Driving Analyzer (PDA) was employed to monitor the hammer performance, pile driving stresses, and soil resistance to driving. Two of the four piles were instrumented with underwater strain gages and accelerometers and the force and velocity signatures for each impact blow were recorded. The maximum hammer Energy Transfer Ratio (ETR) was 50% and the maximum dynamic compression stress measured was 163MPa. The data was further analyzed using signal matching software, CAPWAP, to estimate the driving stresses along the pile, dynamic parameters (quake and damping), and soil resistance distributions along the shaft and the toe. The estimated dynamic parameters and soil resistance to driving are summarized in **Tables 4.6.1-1,** while the capacity from CAPWAP is presented in **Table 4.6.1-2.**

Skirt Pile	Depth of Analysis (m)	EOD/ Restrike	Dynamic Constants			
			Shaft Damping (s/m)	Toe Damping (s/m)	Shaft Quake (mm)	Toe Quake (mm)
A1	80.00	EOD	0.88	1.31	2.60	5.30
A1	80.00	Restrike	0.60	1.31	2.20	2.40
B2	80.00	EOD	0.68	1.44	2.50	8.20

Table 4.6.1-1: CAPWAP estimated dynamic soil parameters

Skirt Pile (Leg)	Vertical Depth of Analysis (m)	EOD /Restrike	Delay Hours	Pile Capacity in Compression (CAPWAP)			
				Shaft Resistance (MN)	Toe Resistance (MN)	Total Resistance (MN)	Soil Setup
A1	80.00	EOD	24.27	11.26	1.74	13.00	2.18*
A1	80.00	Restrike		24.52	2.73	27.26	
B2	80.00	EOD	-	11.54	1.72	13.26	-
(*): Observed soil setup ratio is defined as the ratio of shaft resistance at Restrike to the shaft resistance at EOD of the same pile.							

Table 4.6.1-2: CAPWAP estimated pile capacity

The estimated end-of-driving capacities for A1 and B2 skirt piles were 13MN and 13.26MN, respectively. The restrike test conducted on A1 after 24.27 hrs resulted in a setup of 2.1 with a capacity of 27.26MN. The average end-of-driving capacity was 31% while the restrike capacity was 65% of the expected long-term static capacity of 41.9MN. The project stakeholders would like to know the capacity increase vs. time up so that post-installation activities can be planned.

Long-Term Soil Setup Analyses

Bullock et al (2005) method is used to estimate the long-term setup using the side shear capacities estimated for continuous driving & restrike test from CAPWAP capacities. A side shear semi log-linear setup factor was first computed and then the time required to achieve the required compressive capacity was estimated using the following formula

(Ref) **Table 4.6.1-3.**

$$Q_s/Q_{s0} = (\frac{m_s}{Q_{s0}}) * log(\frac{t}{t_0}) + 1 \qquad \textit{Equation 4.6.1-1}$$

Where,

A = Dimensionless setup factor

t = time elapsed EOD, days

t_0 = reference time, recommended to use 1 day

Q_s = Side shear capacity at time t

Q_{s0} = Side shear capacity at initial reference time t_0

m_s = Semi log-linear slope of Q_s vs log(t) - **(Figure 4.6.1-1)**

In the long run, the pile is expected to be plugged and behave as a closed-end pile. Hence, gross end bearing can be added to the projected shaft resistance for estimating total pile capacity.

A1 SKIRT PILE														
					CAPWAP Capacity				Capacity Estimate with Setup					
	Pile Ver.	Date	Time	Elapsed	Friction	Ratio	Toe	Total	Setup	Friction	Toe	Gross End Bearing	Total (Ru)	
Blow	Penetration	mm/dd/yy		Time (t)	Rs	Rs/Rs0	Rb	Ru	Time	Rs	Rb toe	Rb gross	Rs+Rb toe	Rs+Rb gross
Type	m		hrs	days	MN		MN	MN	days	MN	MN	MN	MN	MN
EOD	80.00	11-11-22	6:03	0.00069	11.25	0.46	1.74	12.99	0.00069	11.3	2.7	6.7	14.0	18.0
Restrike	80.00	12-11-22	6:19	1.01	24.52	1.00	2.73	27.25	1.01	24.5	2.7	6.7	27.3	31.2
(Note: EOD assumes elapsed time of 1 minute)									1	24.5	2.7	6.7	27.2	31.2
									7	28.0	2.7	6.7	30.8	34.8
									14	29.3	2.7	6.7	32.0	36.0
									30	30.7	2.7	6.7	33.4	37.4
									60	32.0	2.7	6.7	34.7	38.7
					End Bearing from Restrike	Rb toe	2.73	MN	90	32.7	2.7	6.7	35.4	39.4
					Plugged End Bearing considering Su 160 Kpa at Pile Toe	Rb gross	6.72	MN	120	33.2	2.7	6.7	36.0	39.9
						t0	1	day	150	33.6	2.7	6.7	36.4	40.4
					Rs at t0	Rs0	24.50	MN	180	34.0	2.7	6.7	36.7	40.7
					Slope of Rs	m	4.20	MN	270	34.7	2.7	6.7	37.4	41.4
					Setup Factor	A	0.17		360	35.2	2.7	6.7	38.0	41.9

Table 4.6.1-3: Projected Compression Capacities w.r.t elapsed time in days (As per Bullock et al 2005) Method

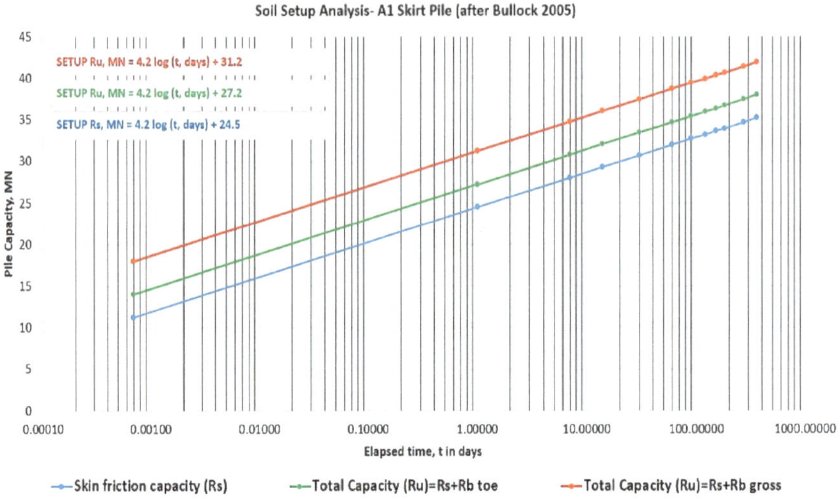

Figure 4.6.1-1: Logarithmic Graph Projections of Pile Capacities w.r.t elapsed time in days

The design static compression capacity of 41.9MN was estimated to be achieved in 360 days from the end of driving.

Conclusions

The soil at this location got remolded due to continuous pile driving resulting in 30% of long-term static pile capacity. The restrike test conducted after 24.27 hrs of end-of-driving measured a soil setup of 2.1 resulting in 65% of long-term pile capacity. These results are in line with Parthasarathy et.al 2015 presented for similar soils.

The Bullock et al (2005) analytical model estimated the time required to achieve a long-term compression capacity of 41.9MN to be 360 days.

Pile monitoring aids in measuring the as-installed pile capacity and the analytical model (Bullock et al, 2005) is useful in estimating long-term setup and the projected pile capacity. This method is generally accepted in the offshore industry.

4.6.2 Case Study – 2 (Dr Prashanth Talkad and Dr Malavika Varma)

Problem Statement

A Jack-up rig may need to be moved nearer to the existing Jacket structure founded on piles. The Jack-up leg penetration causes the soil to be displaced vertically and laterally leading to additional forces and subsequently additional displacements on the existing platform foundation piles. This case study deals with the lateral displacements on a nearby existing Jacket pile resulting from Jack-up leg penetration. FEM analysis was performed to determine if there would be any change in p-y curves resulting in the additional displacement of the existing main piles of a Jacket platform in Offshore Sarawak.

Introduction:

It was planned to mobilize a Jack-up rig near an existing Jacket platform to drive the conductors. The Jack-up spud had a diameter of 16.15m at the widest section. The existing piles have an outer diameter of 1.524m with a penetration depth of 70m. The Jack-up spud geometry and the proximity of the Jack-up with piles location are shown in **Figure 4.6.2-1 and Figure 4.6.2-2.**

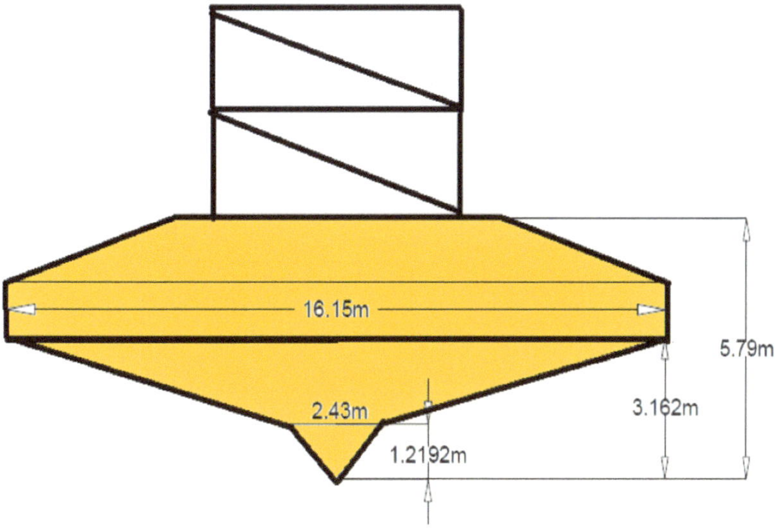

Figure 4.6.2-1: Spud can Geometry

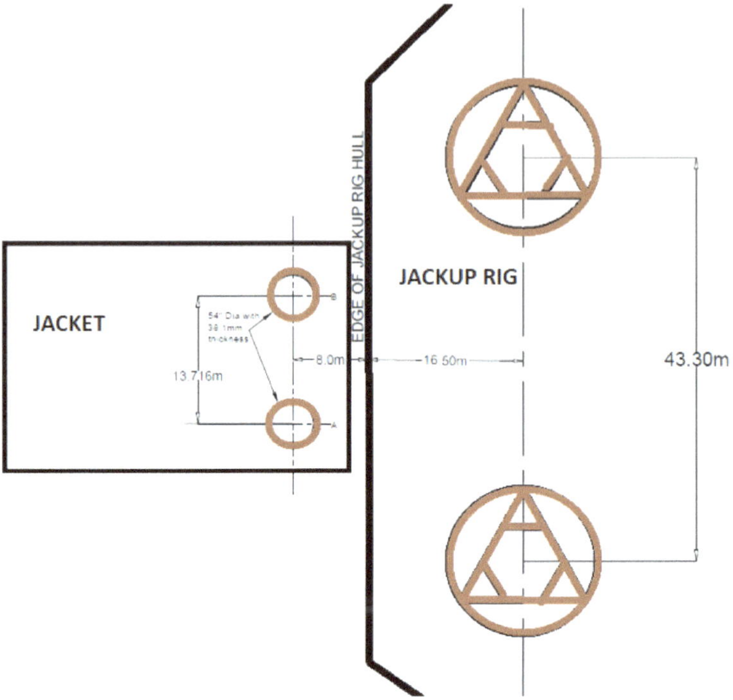

Figure 4.6.2-2: Proximity of the Jack-up rig with Jacket location

Soil Profile

The 150m deep site-specific Geotech borehole indicates alternating Clay and Sand profiles. The seabed soil consists of Sand up to 9.8m followed by stiff Clay up to 16.5m underlain by alternating loose to medium dense Sand and stiff Clay up to the depth of interest of 70m. The soil design parameters considered for Jack-up leg penetration and FEM are summarized in **Table 4.6.2-1.**

Sl. No.	Penetration (m)		Description	Soil Properties				
	From	To		Submerged unit weight, γ' (KN/m3)	Lower Bound Friction angle, φ' (deg)	Upper Bound Friction angle, φ' (deg)	Lower Bound Shear Strength (kPa)	Upper Bound Shear Strength (kPa)
2	3.0	4.6	Dense SAND	10.2	30	35	-	-
3	4.6	10.0	Medium Dense SAND	8.5	25	30	-	-
4	10.0	11.2	Stiff CLAY	9.1	-	-	60	70
5	11.2	14.5	Firm to Stiff CLAY	7.6	-	-	40/50	50/60
6	14.5	16.2	Stiff CLAY	7.6	-	-	60	70
7	16.2	17.8	Medium Dense SAND	8.2	25	30	-	-
8	17.8	20.8	Firm to Stiff CLAY	9.3	-	-	40	50
9	20.8	22	Medium Dense SAND	9	25	30	-	-
10	22	35.8	Stiff CLAY	7	-	-	50/60	60/70
11	35.8	41.8	Medium Dense to Dense SAND	10.7	30	35	-	-
12	41.8	46.0	Stiff to Very Stiff CLAY	7.9	-	-	70/90	90/110
13	46.0	47.0	Stiff CLAY	7.9	-	-	70	90
14	47.0	50.0	Loose to Medium Dense SAND	9	25	30	-	-

Table 4.6.2-1: Soil design parameters

Jack-up Leg Penetration Analysis

The leg penetration analysis was carried out for the derived ground model indicated in **Table 4.6.2.1** and the penetration was estimated to lie between 6.2 m to 7.5 m below the mud line for a maximum preload of 80.1 MN. Therefore, it is unlikely that the legs will punch through the bearing layer. **Figure 4.6.2-3** shows the leg pentation curve.

These predicted penetration values were in close agreement with the observed penetrations when the rig was moved to the location.

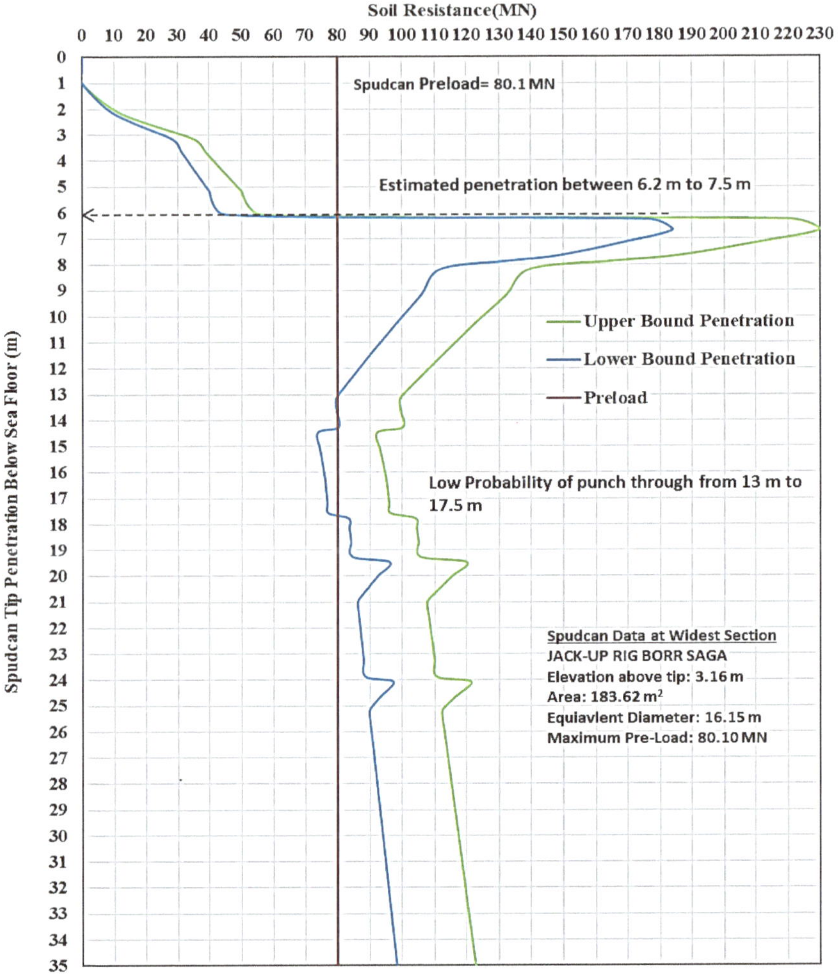

Figure 4.6.2-3: Jack-up Leg Penetration Curve

Finite Element Analysis of Platform Pile and Spud Can Interaction

Problem Geometry

A finite element analysis of platform pile and spudcan penetration uses PLAXIS 3D software. The lateral dimensions of the 3D model are 120 x 120 m, while the model thickness is 73 m with the soil stratigraphy described in **Table 4.6.2-1.** These dimensions were selected such that the model boundaries have negligible effects on the results.

Figure 4.6.2-4 shows the model boundaries in PLAXIS 3D with soil layers. The seabed is assumed to be perfectly flat below the mudline. The finite element mesh comprises of 10-noded tetrahedral elements.

Details of the Finite Element Model

The soil was modelled as an isotropic Elasto-Plastic continuum, with failure described by the Mohr-Coulomb yield criterion. The Clay layers are assumed to behave "undrained" and are characterized by a cohesion equal to the undrained shear strength, Su. The elastic behavior of the clay is defined by a Poisson's ratio of 0.49, and a constant ratio of Young's modulus to undrained shear strength E/Su = 200 is considered in the analysis. The Sand layer is assumed to behave "drained" and is characterized by effective stress shear strength parameters 'c' and 'Φ'.

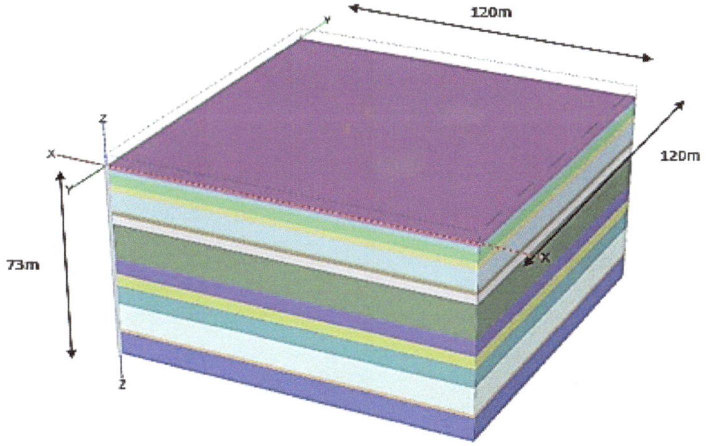

Figure 4.6.2-4: Model Boundaries in Plaxis 3D

The platform pile is modelled as a rigid circular plate with thickness equal to pile wall thickness. The Spudcan is modelled as a weightless elastic (non-porous) rigid body with equivalent width of 16.15m with full base contact. The finite element analysis is carried out for a wished in place analysis with the spud can penetration at 3 m (widest section) below the seabed as shown in **Figure 4.6.2-5.**The displacement contours of the soil model for the 3 m penetration of spud can is shown in **Figure 4.6.2-6.**

The maximum lateral displacement of the pile is 0.009m, while the total soil displacement for the profile is 0.1948m.

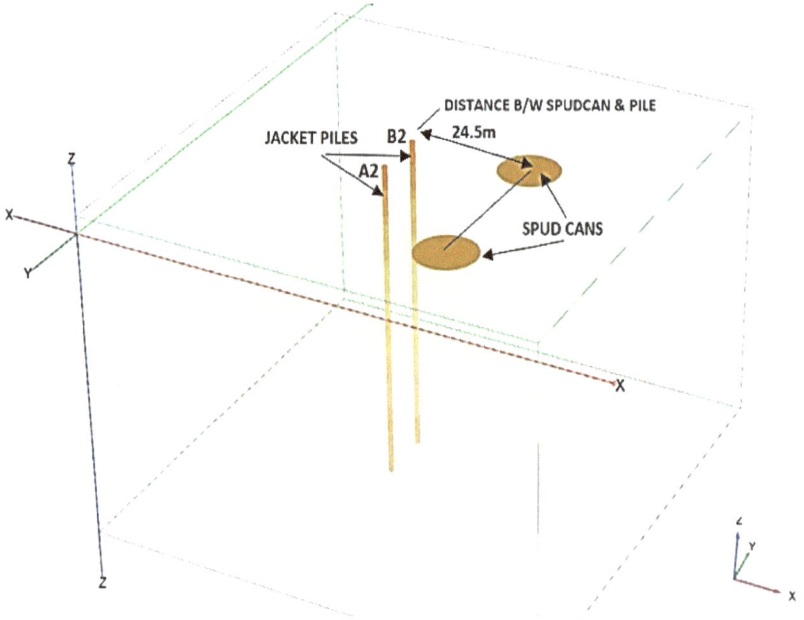

Figure 4.6.2-5: Details of the finite element model in Plaxis 3D

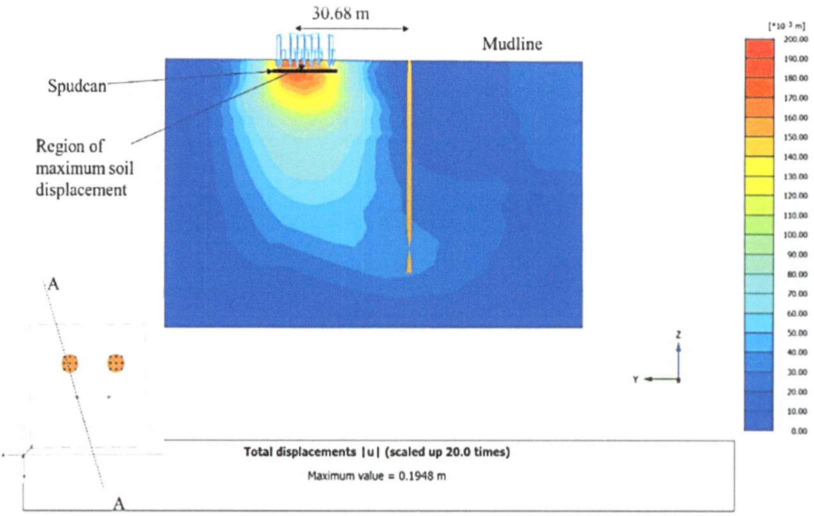

Figure 4.6.2-6: Displacement contours for 3 m penetration of Spud can

Conclusions

The interaction study of the spud can and platform pile was performed using the finite element method (PLAXIS 3D). The spud can was analyzed for the penetration depths of 3m (widest section) and the interference of the same with the platform pile was examined. The lateral deflection and the ground deformation contours were analyzed to understand any possible influence of the spud can penetration on the platform piles. The maximum lateral deflection of the platform pile due to the spud can installation was estimated to be 0.009m. The spud can installation with the wider section resting at a depth of 3m has minimum influence on the platform pile.

The P-y curves for the lateral capacity computation of the platform pile need not be changed because of the negligible lateral soil movement due to spud can penetration.

4.6.3 Case Study – 3 (Mr Kris Gustavian, Gurucharan, Dr Ramesh Gangisetti and Dr CR Parthasarathy)

Problem Statement

Pile driving monitoring is done to measure the hammer performance, dynamic stresses, soil resistance to driving, and pile integrity. Pile integrity refers to a qualitative evaluation of the continuity of pile geometry and the quality of pile material. A pile may have a defect, particularly at splice locations where two small-length pipes are joined by welding either in the fabrication yard or offshore during installation. Therefore, it is imperative to check integrity to plan remedial action if the pile is found defective. This case study deals with evaluating the integrity of a driven pile installed at an offshore location in Malaysia using pile monitoring data.

Project Details:

A caisson pile of 1.524m outer diameter with varying wall thickness between 35 to 60 mm was planned to be driven to its final penetration of 99m at an offshore location in Malaysia. The pile was planned to be driven in three sections, viz., CN-1, CN-2, and CN-3, using an IHC S-280 hammer. Pile monitoring was performed on the last two sections during the driving. The pile has penetrated to 15.5m due to self-weight. The end-of-driving blowcount was 36 blows/0.25 at an ETR (Energy Transfer Ratio) of 79%.

A restrike test was conducted after 53.38 hours with a blowcount of 189 blows/0.25m at an ETR of 79%. The pile monitoring details are summarized in **Table 4.5.3-1** and **Table 4.6.3-2**.

SI. No.	Caisson Section	Hammer	From (in m)	To (in m)
1	CN-2	IHC S-280	15.75	58.50
2	CN-3		59.50	99.00
3	CN-3 Restrike		99.00	99.30

Table 4.6.3-1: Summary of pile monitoring depths

Caisson	Drive	Description	Caisson Section	
			CN-2	CN-3
CN3	BOD	Avg. Energy, kJ	16	140
		Avg. ETR, %	6	50.2
		Blows/0.25m	11	45
		Ver. Pen, m	15.75	59.50
	EOD	Avg. Energy, kJ	99	219
		Avg. ETR, %	36	79
		Blows/0.25m	28	36
		Ver. Pen, m	58.50	99.00
	Restrike/BOD	Avg. Energy, kJ	-	220
		Avg. SE, %	-	79
		Blows/0.25m	-	189
		Ver. Pen, m	-	99.00
EOD: End of drive & BOD: Begin of drive				

Table 4.6.3-2: Summary of pile monitoring parameters

The maximum compression stress at the gage location was calculated by the PDA using the average signal from the two strain transducers. During monitoring, the maximum average compression stresses measured near the gage location during continuous driving of the caisson pile was 121 MPa (35% of steel's yield stress).

During the restrike testing of the caisson pile the maximum stress measured near the gage location was 131 MPa (38% of steel's yield stress). These values are below the API recommended compression stress limit of 310 MPa, or 90% of the steel's 345 MPa yield stress.

Post-Installation CAPWAP Analyses:

CAPWAP analyses have been performed on selected blow records at the end of driving of caisson and a restrike test was conducted on the caisson after a time delay of 53.38 hrs. **Table 4.6.3-3** presents the soil dynamic parameters computed by CAPWAP for the signal matching to derive the caisson capacity. The summary of compression capacities computed by CAPWAP at final penetration is presented in **Table 4.6.3-4.**

Caisson Section	Depth of Analysis (m)	EOD/ BOD/ Restrike	Dynamic Constants			
			Shaft Damping (s/m)	Toe Damping (s/m)	Shaft Quake (mm)	Toe Quake (mm)
CN-3	42.0	EOD	0.554	0.860	1.134	4.257
CN-3	42.0	Restrike	0.600	0.353	1.412	2.377

Table 4.6.3.3: CAPWAP estimated soil dynamic parameters

Caisson Section	Driving Condition	Vertical Pen.	Skin Friction (kN)	End Bearing (kN)	Total Capacity (kN)	Delay (hrs)	Soil Setup Ratio
CN-3	EOD	99.00	5551	875	6426	53.38	3.10
	Restrike	99.00	1722	408	17632		

Table 4.6.3-4: CAPWAP estimated compression capacities during EOD and Restrike

The driving behavior observed in the field was in agreement with the soil stratigraphy. The soil setup ratio observed was 3.10 for a time delay of 53.38 hours at the final penetration of the caisson. The soil resistance to driving (SRD) predicted in the pre-installation studies (Drivability) for continuous driving conditions was found to be higher than the computed/estimated CAPWAP SRD during continuous driving at final penetration.

Caisson monitoring data of CN-2 and CN-3 showed a Beta line at a depth of about 67m from the pile top with Beta values ranging from 77-82% during continuous driving. The force-velocity vs time graph is shown in **Figure 4.6.3-1.**

In general, the Beta line is shown if there is a planned reduction in wall thickness or caisson buckling due to driving. Since the Beta line was observed from the beginning of the drive of CN-2 at a location where wall thickness change was not planned, it was concluded that the caisson pile was defective even before the start of driving. The Beta values further increased due to continuous driving.

It was decided to reject the pile due to a possible reduction in impedance indicating a defect.

Conclusion:

The beta value of 77-86% indicated the possible reduction in the impedance at about 67m below pile monitoring sensors. This beta value was found to decrease (i.e., decrease in the impedance) as the driving progressed, therefore this pile was not recommended to be accepted.

Similar conditions in the Author's experience were encountered in some other offshore projects where the pile was rejected and had to be compensated by installing additional piles.

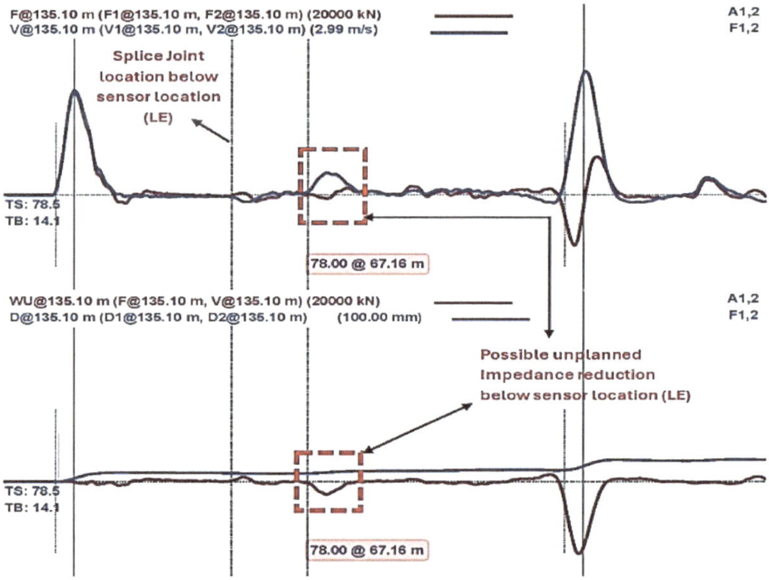

Figure 4.6.3-1: Force-Velocity vs Time graph from PDA

4.7 Associated Risks and Engineering Solutions

The stratigraphy in offshore Malaysia is very similar to that in offshore Indonesia, the mudline is often characterized by the presence of exceptionally soft cohesive materials with shear strengths well below 5 kPa and the stratigraphy of cohesive materials extends to much deeper depths of 150 to 200m below the mudline.

4.7.1 Mud mats

In terms of Mudmat designs, it is important to initially perform high quality geotechnical investigations to acquire good samples to estimate the shear strength of the soil and establish the design profile as accurately as possible to avoid any failure of the mud mats.

4.7.2 Pile Foundations

During driving of pile sections in this region where the presence of deep-seated soft clays is very common, it is important to estimate the pile run-down as accurately as possible based on the geotechnical data and details of the hammer used. Similar to installations in Indonesia, the use of a standard hydraulic hammer generating 280 kJ to 500 kJ of energy is deemed sufficient to drive pile sections to depths ranging from 120m to 130m below the mudline. It is also recommended that the resting stratum of the pile should be able to support the intended load capacity as required. To ensure the achievement of the pile capacity, the conduct of the restrike test shall serve as a critical step of the installation process.

4.7.3 Leg Penetration Analysis for Jack up Rigs

In scenarios where shallow cohesionless deposits are present, intermixed with notable clay seams and desiccated layers, thorough evaluation of leg penetration analysis becomes imperative, particularly to accurately assess punch through conditions. Precise estimates of shear strength parameters for the soil layers, derived from high-quality geotechnical investigations, play a pivotal role in this process. These estimates enable a comprehensive evaluation of Jack up leg penetration, ensuring the rig's stability and safety during operations.

Additionally, consideration of the pullout capacity of Jack up rigs is essential to ensure the safe extraction of the legs, further emphasizing the importance of a meticulous and holistic approach to the study.

4.7.4 Subsea Pipelines

As discussed in previous chapters, the shear strength of the soil strata presented at shallow depths from the mudline have major implications on the design of the pipelines. Based on the estimated subsidence of the pipeline, the pipeline-soil interaction, the effect of scour and corrosion are evaluated in order to ensure the integrity and stability of the pipeline. Therefore, thorough site characterization, zoning of the shallow bed stratigraphy and considerations based on the geophysical data is crucial in the design of the pipeline and the protective coating thickness.

To acquire good samples from the mudline and shallow depths below the mudline stands as a primary requirement for the pipeline-soil interaction studies.

4.7.5 Liquefaction and Scour

Liquefaction and Scour may not have significant effects on the pipelines laid in this region due to the presence of very soft cohesive material at shallow depths from the mudline.

4.8 Lessons from the Experienced

4.8.1 Leadership Journey of Mr Rizal Shah in Offshore Survey: Building a Sustainable Future Through Experience and Ethics

Synopsis

The interview provides valuable insights into the challenges and opportunities in the offshore survey industry, emphasizing the importance of practical experience, safety consciousness, and adaptable business strategies. The interviewee's journey from finance to offshore geotechnical industry leadership demonstrates the value of diverse experiences and continuous learning in building a successful career in this specialized field.

Introduction

In an insightful conversation with a veteran industry leader, Mr Rizal Shah, we explore the fascinating journey of building and leading offshore survey operations across Southeast Asia. From humble beginnings as an accountant to becoming a respected figure in the geotechnical survey industry, this interview provides valuable perspectives on leadership, business resilience, and the critical importance of geotechnical surveys in offshore projects.

Interviewee Profile

The interviewee is a seasoned professional with over 25 years of leadership experience in the offshore survey industry. Starting his career with Shell as an accountant in 1997, he transitioned through various roles including sales and operations management. He currently leads Java Offshore, a prominent survey company operating in Indonesia and Malaysia. His expertise spans business development, operational management, and strategic leadership in the geotechnical survey sector.

The Talk

CRP: Dr. C R Parthasarathy

RS: Mr. Rizal Shah, President, Java Offshore

CRP: Could you share your leadership journey and career progression?

RS: My leadership journey spans about 25 years across different areas. After graduating in 1997, I joined Shell as an accountant. After three years, my boss noticed I was more suited to sales and transferred me.

I spent five years in refined products sales before moving to Brunei, where I managed a diverse team of five people. This experience in Brunei really shaped my understanding of diversity and inclusion. In 2005, I moved to Indonesia to start up Shell's commercial business, and in 2011, I became CEO for Offshore Works Indonesia (OWI).

CRP: What were some significant milestones or challenges in your journey?

RS: Two events particularly stand out. First was achieving our $40 million revenue target within five years of starting the company. The second, more sobering experience, was in April 2018 when we had a vessel catch fire due to a shallow gas blowout. This incident, which resulted in one fatality, was a significant wake-up call about safety in our operations.

CRP: How critical is geotechnical survey work in oil and gas projects?

RS: It's as critical as breathing air. While procurement often focuses on big-ticket items worth billions, our work, though relatively smaller in cost, determines whether you'll spend one billion or three billion on a project.

Unfortunately, there isn't enough recognition for its importance, and many midsized and some large oil companies do not even have their own geotechnical specialists.

CRP: What's your business philosophy and approach to staying competitive?

RS: Our business model focuses on a 'fit for purpose' model. We cater our services based on what the client needs to ensure a proper balance between sustainability and consistent results. The challenge in our industry, especially within the segments we operate, is the dynamics related to commodity prices. Ensuring a prudent approach to the business will differentiate our ability to adapt to changes in market volatility.

CRP: What's your vision for the future of the company?

RS: My vision for the company is to establish it as a recognized leader in our specialized industry—known for our capabilities, reputation, and commitment to sustainability. Realistically, I aim to make us the best in this region, creating a strong foundation of responsible professionals who can drive innovation and sustainable development.

While global dominance like Fugro or Google isn't achievable in my lifetime, I believe my successor can take us there. For now, it's about building a brand that can sustain itself and grow over time. Success isn't just about size or revenue; it's about creating lasting value and contributing meaningfully to the industry.

Looking back, I'm proud of how far we've come despite challenges—there were times when even running operations felt impossible. But every low point taught me resilience and reminded me that opportunities always exist if you're willing to push forward. This journey has been humbling, and I feel blessed to have achieved what once seemed unimaginable.

CRP: What's your outlook for the industry over the next 12 months and five years?

RS: The next 12 months look positive with growth expected across all segments: Oil & gas, infrastructure, telecommunications, and renewables. Currently, all our assets are fully occupied. However, the five-year outlook is harder to predict. While renewable energy markets are expected to grow by 30% per annum, oil and gas will grow at a slower pace but still maintain about 70-80% market share of global energy needs.

CRP: How do you manage work-life balance and stress?

RS: I allocate time for exercise at least three times a week, whether it's running, cycling, or bowling. Being in a 24/7 industry means I can't completely disconnect, but I maintain a structured routine of checking emails and approvals in the morning before spending time with people. I'm fortunate to have a very understanding wife who has supported my career journey and managed our family while I traveled extensively.

4.8.2 Leading Through Innovation: McDermott's Century of Excellence in Energy Solutions - A Conversation with Vice President of Operations, Mr. Biswadeep Das.

Synopsis

This interview provides valuable insights into McDermott's successful journey through a century of energy industry evolution. The company's commitment to core values, operational excellence, and sustainable future development, combined with its robust project portfolio and market position, suggests a promising trajectory for the next century. The emphasis on transparency, innovation, and sustainable solutions demonstrates McDermott's readiness to address future energy challenges while maintaining its position as an industry leader.

Introduction

As McDermott International celebrates its centennial milestone, this exclusive interview provides unique insights into the company's evolution from its humble beginnings in Texas to its current status as a global energy solutions provider.

Through the lens of a seasoned industry veteran, we explore McDermott's core values, operational excellence, and vision for sustainable energy solutions in an ever-evolving industry landscape.

Interviewee Profile

The interviewee is a Vice President of Operations at McDermott International, with 29 years of industry experience. Starting as a field engineer on offshore installation vessels, he has progressed through various roles including procurement, fabrication management, engineering management, and project management. Currently, he oversees all projects executed from Asia and is responsible for business acquisition in the region. His six-year tenure at McDermott has been marked by progressive leadership roles, from Project Director to his current position as Vice President. His extensive experience spans multiple geographies including India, the Middle East, Mexico, and Malaysia, where he has been based for the last decade.

The Talk

CRP: Could you provide an overview of McDermott's history and current operations?

BD: McDermott was founded in 1923 by Ralph McDermott, initially building wooden drill rigs in Texas. Today, after celebrating its centennial in October 2023, the company operates in 56 countries with over 35,000 employees. McDermott serves the energy industry comprehensively, including low-carbon solutions, upstream oil and gas, tanks, and specialized LNG solutions.

CRP: What are McDermott's core values and operational principles?

BD: McDermott operates on five core values: integrity, one team approach, commitment, well-being of all stakeholders, and going beyond. Additionally, the company follows 12 Operational Values (MOVs) that guide project lifecycle management, including crucial aspects like readiness planning and mechanical lifting. These values and operational principles are consistently applied across all global operations.

CRP: Can you share some significant challenges and lessons learned from your project experience?

BD: One notable challenge occurred in 2005 during the RS2 platform project in Bombay High, where all four piles encountered refusal due to unexpected soil conditions.

The solution involved importing specialized high-strength grout from Denmark. This experience taught the importance of preparing for worst-case scenarios and increasing underdrive allowance. Another valuable lesson came from a Qatar project where inadequate crane boom sea fastening led to significant complications, emphasizing the importance of attention to both major and minor details in project execution.

CRP: What advice would you give to the younger generation in the industry?

BD: Two key pieces of advice: First, never hide problems - earlier identification leads to easier solutions. Second, no idea to solve a problem should be dismissed as bad. Even seemingly unconventional solutions, like the one found for the pile refusal issue through research of similar cases, can prove valuable.

CRP: Were there any role models or mentors who significantly influenced your life and career

BD: My primary inspiration is Dhirubhai Ambani, for his impact on India's growth.

Professionally, I'm indebted to two mentors: Mr. D.K. Gupta of L&T, who provided invaluable guidance early in my career, and Mr. Mahesh Swaminathan, whose drive and innovative thinking continue to inspire me.

CRP: What is McDermott's vision for the future?

BD: McDermott is committed to maintaining its legacy of reliable project delivery while focusing on sustainability and Net Zero goals for 2050. The company aims to make energy more affordable, particularly in developing regions, and is actively contributing to the transition from fossil fuels to gas-based economies in countries like Vietnam.

CRP: What is the current market outlook for McDermott?

BD: McDermott currently has its highest-ever backlog, exceeding $20 billion. Under strong leadership, the company sees positive growth potential not just for the next five years but for the next decade and beyond, particularly in green energy initiatives.

CRP: Dr. C R Parthasarathy

BD: Biswdeep Das, Vice President- Operations, McDermott

4.8.3 Breaking Barriers in Oil & Gas: A Female Engineer's Journey to Leadership at Sapura Energy - Ms. Linda.

Synopsis

Ms. Linda's journey from offshore operations to engineering leadership exemplifies how technical expertise, adaptability, and resilience can overcome industry barriers. Her emphasis on continuous learning, innovation, and thorough verification processes provides valuable insights for future engineers. As Sapura Energy looks to expand into new markets, particularly floating facilities, her leadership approach combining technical excellence with strategic vision positions the company for continued growth in the evolving oil and gas sector.

Introduction

In an inspiring conversation that spans nearly three decades of experience in the oil and gas industry, Ms. Linda shares her remarkable journey from being one of the few women in offshore operations to becoming the Head of Facilities Engineering at Sapura Energy. Her story exemplifies how technical expertise, resilience, and innovative thinking can break gender barriers and drive success in a traditionally male-dominated industry.

Interviewee Profile

Ms. Linda is the Head of Facilities Engineering at Sapura Energy, bringing 27 years of experience in the oil and gas industry.

Her career trajectory includes 13 years in operations and maintenance on the client side before transitioning to contractor services. A Fellow of the Institute of Mechanical Engineers (UK) at an unusually young age, she has led significant projects across multiple geographies, including Africa, Gulf of Mexico, Australia, and Brazil. Her expertise spans FPSO operations, engineering design, and project management, with notable achievements including a 15% boost in production reliability that earned her a global production recognition award.

The Talk

CRP: Dr CR Parthasarathy

ML: Ms. Linda, Head of Facilities Engineering at Sapura Energy

CRP: Could you elaborate on your journey to leadership at Sapura Energy?

ML: Before joining Sapura Energy, I spent 13 years on the client side, focusing on operations and maintenance. The hands-on experience in sustaining production led me to transition to the contractor side. I initially joined Sapura Energy's Project Management Unit (PMU) for risk assessment and project performance monitoring, before being promoted to head the facilities engineering team due to my extensive engineering experience.

CRP: What significant experiences shaped your career?

ML: One defining moment was working on an FPSO in the Equator Region over 20 years ago as the only female engineer. Despite initial gender bias, I successfully improved production reliability by 15%, earning a global production recognition ***'Helping Hands'*** award. This success led to regional roles and eventually caught the attention of contractor companies.

CRP: What are Sapura Energy's core philosophy and principles?

ML: Sapura Energy follows 'Our Way,' which comprises three values - Honest, Accountable, and Respectful - and three principles: Commitment to Collaboration, Spirit of Innovation, and Pride in Results. These principles drive cross-functional teamwork between engineering and operations while fostering innovative solutions without compromising quality.

CRP: What challenges have you faced and what lessons were learned?

ML: A significant lesson came from a near-disastrous incident involving incorrect geotechnical input data. Despite having senior personnel involved, human error in data verification led to serious consequences. This experience emphasized the critical importance of thorough data verification, regardless of the expertise level of those involved.

CRP: What advice would you give to young engineers, especially women?

ML: I encourage women to venture into challenging disciplines like geotechnical and machinery engineering. Learn from experienced engineers, seek hands-on experience, and don't let rejection or cynical remarks discourage you. Use challenges as opportunities to prove your capabilities and contribute to the company's success.

CRP: What's your vision for Sapura Energy's future?

ML: Sapura Energy should expand beyond its current strengths in T&I and EPC contracting into the floating facilities business, particularly FPSO operations. Given our vessel ownership and topside engineering expertise, we could partner with vessel owners initially and potentially become co-owners of FPSOs later.

CRP: What is the market outlook for Sapura Energy in the next 12 months, and how do you foresee the oil and gas market evolving in the region over the next five years?

ML: In the next 12 months, Sapura Energy is projected to maintain a strong position, particularly in the TNI (Transportation and Installation) sector. This is due to our specialized expertise, robust vessel fleet, and highly competent engineering, naval, and geotechnical personnel.

We possess extensive experience operating in diverse and challenging environments globally, including South Africa, other parts of Africa, the Gulf of Mexico, Australia, and Brazil, in addition to Southeast Asia. This global experience, combined with our proven track record, solidifies our position as a leading TNI contractor.

Looking forward five years, we anticipate continued demand for specialized services like ours, even as the energy landscape shifts. We expect to see continued growth in projects requiring our unique expertise.

CRP: Could you share your daily habits, hobbies, and a bit about your family?

ML: Reaching my fifties has been a privilege, prompting a shift to hobbies like gardening and reading historical biographies, which bring tranquility and inspiration. Travel remains a passion, offering cultural insights and quality time with my family, especially now that I have more time at home. I have four grown children: two in marine engineering, one in STEM (Science, technology, Engineering and Mathematics) as a computer programmer, and two still in secondary education, and I cherish the opportunity to spend more time with them after years of demanding work.

PS: Sapura Energy Berhad changed its name to Vantris Energy Berhad on 01 August 2025.

5
ABU DHABI

5.1 Offshore Abu Dhabi: Geology and Geotechnical Engineering Challenges

The geology of offshore Abu Dhabi is characterized by a layered sequence of carbonate reservoirs, structural traps, and complex near-surface sediments. These geological features are critical for understanding hydrocarbon systems, subsurface stability, drilling challenges, and infrastructure design in the region.

5.1.1 Stratigraphy and Reservoir Geology

The subsurface geology of offshore Abu Dhabi is dominated by Jurassic to Cretaceous carbonate formations, which host major hydrocarbon reservoirs (Grötsch et al. 2003). Below is a summary of key formations:

1. Arab Formation (Upper Jurassic): The Arab Formation consists of dolomitized limestones and anhydrite seals, forming prolific oil reservoirs. The Hith Anhydrite, part of this formation, acts as a regional seal controlling oil distribution. Notably, these reservoirs contain sour gas with hydrogen sulfide (H_2S) concentrations reaching up to 33%, requiring corrosion-resistant materials for drilling and infrastructure.
2. Thamama Group (Lower Cretaceous): This group includes layered limestone reservoirs such as the Kharaib, Shuaiba, and Lekhwair members. These reservoirs exhibit varying porosity and permeability profiles, making them heterogeneous but highly productive. The Thamama Group straddles the eastern edge of the Hith Anhydrite seal.
3. Mishrif Formation (Middle Cretaceous): The Mishrif Formation contains minor oil-bearing carbonate reservoirs with rudist-rich facies.

While more significant onshore, it contributes to hydrocarbon production in some offshore fields.

4. Araej Formation (Middle Jurassic): This formation hosts gas and condensate-bearing limestones. Deep-seated faults influence its distribution, posing risks of unexpected gas pockets during drilling operations.

5.1.2 Structural Controls

The Hith Anhydrite seal creates a north-south trending boundary that divides the offshore area into two distinct zones:

1. West of the Hith Edge: Dominated by Arab Formation reservoirs sealed by anhydrite.
2. East of the Hith Edge: Thamama reservoirs dominate, with fault-assisted migration pathways facilitating hydrocarbon movement.

Deep-seated faults play a dual role:

1. They facilitate vertical hydrocarbon migration, increasing the risk of encountering unexpected gas pockets during drilling.
2. Fault reactivation can lead to reservoir compartmentalization, affecting fluid flow and pressure regimes.

Understanding these structural controls is essential for predicting drilling hazards and optimizing well placement.

5.1.3 Source Rocks and Hydrocarbon Generation

Two primary source rocks drive the hydrocarbon systems in offshore Abu Dhabi: (Grötsch et al. 2003)

1. Diyab Formation (Upper Jurassic): Composed of argillaceous limestones, the Diyab Formation is a prolific generator of hydrocarbons for both Arab and Thamama reservoirs. Its high organic richness and thermal maturity make it a key contributor to the region’s oil and gas resources.
2. Shilaif Formation (Middle Cretaceous): This organic-rich carbonate formation sources hydrocarbons for the Mishrif reservoirs. It is particularly important in the western part of Abu Dhabi, both onshore and offshore.

Thermal maturity modeling is critical for understanding hydrocarbon generation timing and predicting pore pressures, which directly impact drilling safety and efficiency.

5.1.4 Near-Surface Geology

The shallow subsurface (<25 m) comprises unconsolidated sediments that pose unique geotechnical challenges:

1. Abu Dhabi Formation (Holocene): This formation includes carbonate sands, algal mats, and sabkha evaporites.
   These unconsolidated sediments require careful foundation design for offshore structures. Sabkha zones are prone to salt weathering and subsidence, necessitating reinforced pilings and corrosion-resistant materials.
2. Ghayathi Formation (Pleistocene): Cemented dune sands from this formation form core layers in barrier islands, providing some stability to shallow foundations.

Sabkha-related corrosion is a significant concern due to chloride-induced degradation of steel pilings in marine environments. Engineers must account for these factors when designing offshore infrastructure.

5.1.5 Exploration Technologies and Data

Recent advancements in exploration technologies have significantly improved our understanding of offshore Abu Dhabi's subsurface:

1. 3D Mega Seismic Surveys: These surveys map reservoirs at depths of up to 7,600 meters, resolving fault networks and fluid contacts. Such high-resolution data is essential for identifying drilling hazards, such as over pressured zones, and optimizing well paths.
2. Integrated Geological Models: Combining sequence stratigraphy, diagenesis, and dynamic reservoir behavior, these models provide a comprehensive understanding of reservoir properties and fluid dynamics. They are indispensable for mitigating risks during drilling and infrastructure placement.

Emerging technologies, such as machine learning and artificial intelligence, are increasingly being used to interpret seismic data and predict reservoir characteristics with greater accuracy.

5.1.6 Key Considerations for Geotechnical Engineers

For geotechnical engineers working in offshore Abu Dhabi, several critical factors must be addressed:

1. Managing H_2S Risks: The presence of sour gas (up to 33% H_2S) in deep reservoirs requires the use of corrosion-resistant materials in drilling equipment and infrastructure.
2. Mitigating Sabkha-Related Corrosion: Sabkha zones are prone to chloride-induced corrosion of steel pilings. Reinforced pilings and specialized coatings are necessary to ensure long-term structural integrity.
3. Avoiding Fault Zones: High-resolution subsurface models help engineers avoid fault zones during infrastructure placement, reducing the risk of seal breaches and fluid escape.
4. Foundation Design in Unconsolidated Sediments: The unconsolidated nature of near-surface sediments necessitates robust foundation designs, often involving reinforced pilings and other stabilization techniques.

Conclusion

Offshore Abu Dhabi's geology is defined by a complex interplay of carbonate reservoirs, structural traps, and near-surface sediments. Understanding these features is crucial for addressing the challenges of subsurface stability, drilling safety, and infrastructure design. By leveraging advanced exploration technologies and integrated geological models, engineers can mitigate risks and optimize resource extraction in this hydrocarbon-rich region.

This overview provides a foundation for engineering students and the general public to appreciate the geological and geotechnical complexities of offshore Abu Dhabi while highlighting the importance of accurate data and innovative solutions in addressing these challenges.

5.2 Introduction

Abu Dhabi, the capital of the United Arab Emirates (UAE), is renowned for its vast hydrocarbon reserves, which have been pivotal in driving the emirate's economic growth and development. Situated along the Arabian Gulf coast, Abu Dhabi's offshore fields play a crucial role in the nation's oil and gas industry, contributing significantly to both domestic energy supply and global energy markets. The UAE is among the world's ten largest oil producers.

The offshore fields of Abu Dhabi are characterized by their substantial reserves, advanced production techniques, and strategic importance in the global energy landscape, Ref: (**Figure 5.2-1**).

These fields are managed and operated primarily by the Abu Dhabi National Oil Company (ADNOC) and its subsidiaries, with partnerships formed with leading international oil companies to leverage expertise and technology.

The U.S. Energy Information Agency cites figures estimating the UAE holds the seventh-largest proven reserves of natural gas in the world at over 215 trillion cubic feet (International Trade Administration 2023).

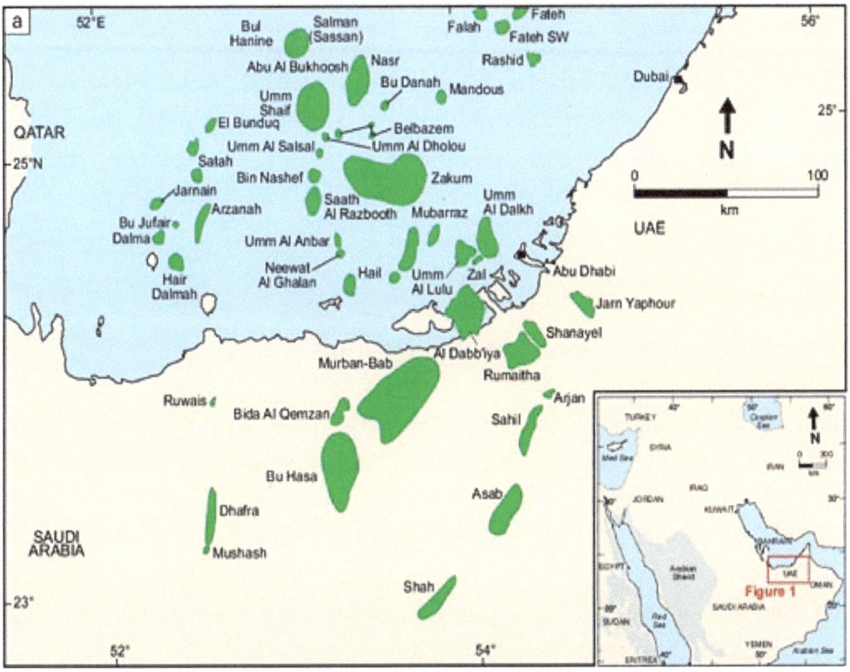

Fig 5.2-1: Location Map of Onshore & Offshore field of Abu Dhabi (Morad et al, 2012)

In February 2022, ADNOC announced the discovery of natural gas resources offshore of Abu Dhabi. This discovery marks the first from Abu Dhabi's offshore exploration concessions, highlighting the continued success of ADNOC's block bid rounds and its expanded approach to strategic partnerships. The country is seeking to become self-sufficient in gas supply by 2030. (Ref: http://surl.li/smykw - International Trade Administration, USA)

5.2.1 Notable Offshore Fields:

- **Zakum Field**: Zakum is one of the largest and oldest offshore oil fields globally, located approximately 84 kilometers northwest of Abu Dhabi. It has been a cornerstone of the emirate's oil production since the 1960s, with reservoirs like the Upper Zakum developed using state-of-the-art drilling and production technologies.
- **Umm Shaif and Lower Zakum Fields**: Adjacent to Zakum, these fields contribute significantly to Abu Dhabi's oil production capacity. Operated by ADNOC Offshore, they employ advanced techniques to maximize hydrocarbon recovery while meeting stringent environmental standards.
- **Satah Al Razboot (SARB) and Umm Al Dalkh Fields**: These newer offshore assets, located approximately 120 kilometers northwest of Abu Dhabi, utilize cutting-edge technologies for enhanced oil recovery, including artificial islands and advanced drilling methods. They reflect Abu Dhabi's commitment to innovation and sustainable resource management.

5.3 General Geology of the Region

The emirate of Abu Dhabi has a total area of approximately 107,700 sq. km, comprising 77,700 sq. km of onshore land and 30,000 sq. km of offshore territory in the Persian Gulf. This constitutes the vast majority of the UAE's total area of 84,000 sq. km (The Free Library 2025).

Offshore acreages with rich sedimentation have reached about 30,000 sq.km. Water depths off Abu Dhabi seldom exceed 40 meters. In terms of tectonic framework, the UAE is situated within the interior platform of the Arabian shelf, bounded on the NW by the Qatar - South Fars Arch, and on the east and NE by the foreland basin, and adjacent foreland fold and thrust belt, of Oman. Two passive plate margin basins were developed in the area - the Rub Al Khali and Ras Al Khaimah Basins.

The Permian - Recent sequence here is dominated by marine carbonates, with occurrences of terrigenous clastics and evaporites associated with cyclic periods of marine regression. Major oil and gas reserves have been discovered in the emirate since the 1950s. It was identified that the area of highest residual source-rock potential is in western onshore and offshore Abu Dhabi and that the source potential ranges from moderate to good.

Deep wells drilled have penetrated the pre-Khuff clastics of Permian and pre-Permian age. Lower Paleozoic salt, possibly Cambrian in age, is believed to underline much of Abu Dhabi, especially offshore. Salt structures like those exposed to the north in Iran are believed to have been responsible for much of the structuration within the basin, uplifting younger rocks to form the traps for hydrocarbons.

Abu Dhabi's proven oil and gas reserves are 97.8 Bbbl and 215 Tcf (trillion cubic feet), respectively. It produces 3.471 Mbbl/day (thousand barrels per day), which is ranked 6th in the world (Jaehong Ko 2015). Since the Permian, UAE was in low latitude by the Equator and became the depositional site of carbonate that consists of excellent reservoirs.

Especially Cretaceous Thamama Group which includes rudist buildup complexes is the most petroliferous and constitutes several offshore- and onshore giant oil fields such as Zakum, Umm Shaif, Bu Hasa, Asab, Bab and Sahil. Oil was generated from argillaceous lime mudstones rich in organic matter deposited in anoxic intrashelf basins including Diyab, Bab, and Shilaif. Hith evaporite, and shales and fine-grained carbonate rocks of the Nahr Umr, Laffan, Fiqa, and Umm Er Radhuma formations act as seals for vertical migration and influence on distribution of oil and gas accumulations. Anticlinal traps were developed involving basement movement or salt tectonics.

5.3.1 General Stratigraphies

The following section provides a basic understanding of the typical stratigraphy (Ref: **Table 5.3.1-1 and Table 5.3.1-2**) of some of the most important offshore oil and gas fields in Abu Dhabi.

Nasr Field

Depth (m)	Description	Shear Strength (Su), kPa	UCS of Rock (MPa)	Angle of Internal Friction (phi)
0 – 23	Moderately Weak Siliceous CALCERNITE	-	2.0 – 7.5	-
23 – 24	Moderately Weak Siliceous CALCISILTITE	-	1.0 – 2.0	-
24 – 29	Loose to Medium Dense Siliceous Carbonate SAND	-	-	28
29 – 35	Very Stiff Siliceous Carbonate Sandy SILT	150 – 200	-	-
35 – 64	Loose to Medium Dense Siliceous Carbonate Silty SAND / Sandy SILT		-	30
64 – 65	Moderately weak Clayey CALCILUTITE	-	1.0 – 2.0	-
65 – 67.5	Moderately weak Siliceous CALCISILTITE	-	1.0 – 2.5	-
67.5 – 75	Very Stiff Siliceous Carbonate Sandy SILT	175 – 250	-	-

Table 5.3.1-1: Generalized Soil Stratigraphy of Nasr Field

Umm Shaif

Depth (m)	Description	Shear Strength, kPa (S_u)	UCS of Rock (MPa)	Angle of Internal Friction (phi)
0 – 7.5	Medium Dense Siliceous Carbonate SAND	-	-	27
7.5 – 10	Moderately Weak Siliceous CALCERNITE	-	1.0 - 7.5	-
10 – 12.5	Stiff Siliceous Carbonate Clayey SILT	50 – 100	-	-
12.5 – 30	Moderately Weak Siliceous CALCERNITE	-	2.0 – 10.0	-
30 – 55	Very Stiff Siliceous Carbonate Clayey SILT	120 – 200	-	-
55 – 70	Very Stiff to Hard Siliceous Carbonate Silty CLAY	100 – 200	-	-
70 – 72	Dense Siliceous Carbonate Silty SAND	-	-	35
72 – 75	Hard Siliceous Carbonate Silty CLAY	150 – 200	-	-
75 – 77	Dense Siliceous Carbonate Silty SAND	-	-	35
77 – 82	Moderately Weak Siliceous CALCERNITE	-	2.0 – 7.5	-
82 – 86	Very Stiff Siliceous Carbonate Sandy SILT	125 – 200	-	-
86 – 90	Moderately Weak Siliceous CALCERNITE	-	2.0 – 5.0	-

Table 5.3.1-2: Generalized Soil Stratigraphy of Umm Shaif

5.4.1 Case Study – 1 Evaluation of Load transfer vs Deflection curves due to soil liquefaction (Dr Prashanth Talkad and DR Dr Parthasarathy)

Problem statement

The liquefaction susceptibility is evaluated by considering the soil properties alone, without considering the earthquake loading. If soil at a particular site is susceptible to liquefaction, then only it is prone to liquefaction hazards. The important factors that will decide the susceptibility of soil liquefaction at the site are soil type, soil state, particle shape and soil permeability. It is required that the soil strengths be degraded for estimating the longitudinal and lateral load-deformation characteristics- P-Y, T-Z and Q-Z-if the site is susceptible to liquefaction. This case study discusses the methods available for obtaining degraded P-Y for liquefied soils and estimation of the same.

Project Details

A Single Point Mooring (SPM) system was to be installed in offshore Middle East. The geotechnical investigation program consisted of seven deep boreholes with final depths ranging from 25m to 53m. Stratigraphy in general consists of Very Soft SILT/Silty CLAY in the top 15m, followed by medium dense Silty SAND up to about 25m to 30m. Below which, alternate layers of dense to very dense Silty Sand and Stiff Clay layers are encountered up to the investigated depth. The liquefaction susceptibility of the region was analyzed based on grain size distribution, index properties, and cone penetration test (CPT) results.

The analysis showed that the soil strata within the top 3m was susceptible to liquefaction under earthquake loading. It was required to develop the load transfer curves (P-Y, T-Z and Q-Z) due to soil liquefaction based on the Beam on Nonlinear Winkler Foundation Model for 3 different sizes of piles viz., 36", 42" and 48" OD with a wall thickness of 25mm for the installation of Single Point Mooring.

P-Y models

The pile foundation should be designed to sustain lateral loads, whether static or cyclic. The lateral resistance of the soil near the surface is significant to pile design and the effects on this resistance, scour and soil disturbance during pile installation should be considered.

Generally, under lateral loading, clay soils behave as a plastic material which makes it necessary to relate pile soil deformation to soil resistance. To facilitate this procedure, lateral soil resistance deflection (p-y) curves should be constructed using stress-strain data from laboratory soil samples.

The representation of lateral soil resistance is done using p-y springs in a Beam-on-Non-Linear Winkler foundation method (BNWF) to evaluate both the inertial and the kinematic pile response in liquefied and/or laterally spreading soil. The BNWF model for normal soil is widely practiced but for liquefied soils the interaction between pile and soil is much more complicated. Hence, a reduction factor in the soil resistance "p" is used which is dependent on conditions such as degree of liquefaction, depth of liquefied soil, etc. Some of the most commonly used p-y curve models are p-multipliers, Cu factor, Residual shear strength method, Rollins method and Hybrid p-y model. The reliability of these methods for representing lateral resistance of liquefied soil may simply be limited by the fact that they are approximations of a rather complex phenomenon that is poorly understood.

p-multipliers

The influence of liquefaction on BNWF springs for sand can be approximately accounted for by applying scaling factors, or p-multipliers (m_p), to the p-y resistances (Scott Ashford et.al 2011). **Table 5.4.1-3** summarizes the p-multipliers vs SPT values 'N'.

SPT, N	P-multiplier
<8	0.0 to 0.1
8-16	0.05 to 0.2
16-24	0.1 to 0.3
>24	0.2 to 0.5

Table 5.4.1-3 : P-multipliers, m_p, to account for liquefaction

C_u Factor

Instead of a P-multiplier for a fully liquefied soil, Cu Factor method assumes strength degradation factor according to the degree of liquefaction. From centrifuge model tests, Liu and Dobry (1995) have derived the Cu as a linearly varying function with excess pore pressure ratio (i.e., Cu = 1-ru).

Residual shear strength method

Wang and Reese (1998) developed a different approach to predict p-y curves for pile foundations in liquefied soil. Recognizing that the residual shear strength of the liquefied soil should limit the soil resistance at larger strains, they recommended using the p-y relationships developed by Matlock (1970) for soft clay in conjunction with the predicted residual shear strength of the liquefied soil. The residual shear strength of liquefied soils is commonly estimated in practice using residual strength ratio methods (Olson and Stark 2002; Idriss and Boulanger 2007; Ledezma and Bray 2010) and/or uniform residual strength methods (Seed and Harder 1990), both of which are typically functions of $(N1)_{60}$.

Rollins Method

Ashford and Rollins (2002) performed full-scale load tests on piles in the field using blast-induced liquefaction techniques to study the inertial load-displacement behavior of piles in liquefied medium dense sands. Their tests showed a dilation effect in the liquefied soil that caused the observed p-y behavior to have a concave-up shape. These curve shapes were also observed in subsequent full-scale tests on larger drilled shafts (Rollins et al. 2005b). Similar dilation effects had previously been reported by other researchers such as Wilson et al. (2000). Rollins et al. (2005a) later developed an empirical mathematical model to quantify the dilative p-y behavior of the liquefied sand observed in their field tests. This mathematical model is given as

$$p = A(By)^c(p_d) \qquad \textit{Equation 5.4.1-1}$$

$$A = 3 * 10^{-7}(z+1)^{6.05} \qquad \textit{Equation 5.4.1-1}$$

$$B = 2.80(z+1)^{0.11} \qquad \textit{Equation 5.4.1-2}$$

$$C = 2.85(z+1)^{-0.41} \qquad \textit{Equation 5.4.1-3}$$

$$p_d = 3.81(ln\ ln\ d) + 5.6 \qquad \textit{Equation 5.4.1-4}$$

Where,

p = soil pressure per unit length of the pile (kN/m)

y =relative differential displacement between the pile and soil (mm)

z = depth of interest from the ground surface (m)

pd = diameter adjustment factor, and

d = diameter or width of the pile (m).

Hybrid p-y model

Hybrid p-y model combines existing p-y relationship based on residual shear strength method and Rollins method to form a simple hybrid method. The proposed hybrid p-y model simply requires selecting the lower predicted soil resistance, (Pi) min, from the Wang and Reese (1998) and Rollins et al. (2005a) p-y models for each relative differential displacement between the pile and soil, yi. In the hybrid p-y model, it is recommended that the residual shear strength of the liquefied soils be estimated using the 33rd-percentile values from the Seed and Harder (1990) relationship. It is recommended that the residual shear strength of the liquefied soils be set to zero for soils with $(N1)_{60}$ = 5 blows per foot (i.e., Dr ,30%), which has been field validated for large strains in very loose liquefied soils by Juirnarongrit and Ashford (2006). Seed and Harder (1990) also truncate their residual shear strengths at a value of $(N1)_{60}$, which corresponds to an approximate relative density of 55%. Recommended residual shear strengths based on Seed and Harder (1990) are shown in **Figure 5.4.1-1.**

Franke & Rollins (2013) found that with a ε50-parameter set at 0.05, hybrid p-y models yielded reasonable results in very loose to loose liquefiable soils. However, for medium-dense liquefiable soils, errors were significant due to lower strain dilation, making actual soil resistance stiffer. They suggested modifying ε50 as a function of $(N1)_{60}$.

$$0.0005 \leq \varepsilon_{50} = 0.1537e^{-0.229[(N_1)_{60-cs}]} \leq 0.05 \quad \textit{Equation 5.4.1-6}$$

As a sample basis, hybrid model is developed for one borehole, where the stratigraphy primarily consists of loose Silty Sand in the top 6.0m.

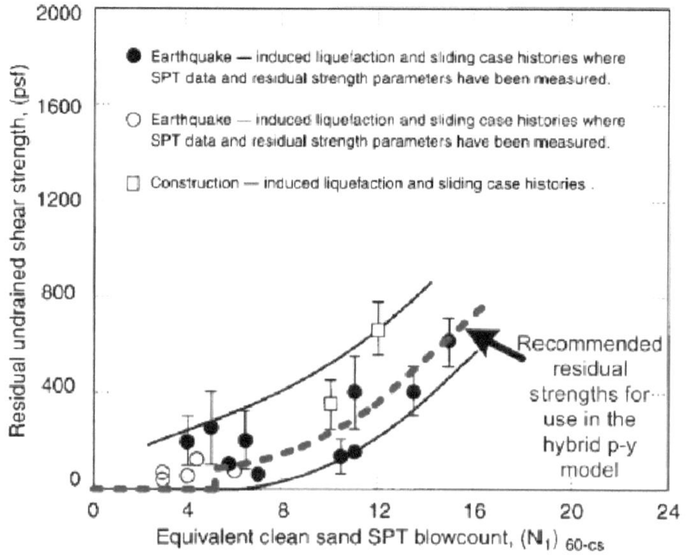

Figure 5.4.1-1: Recommended method for estimating residual shear strength

Based on CPT data, the (N1)60 value is estimated to be in the order of about eight. The corresponding residual shear strength is about 9 kPa. **Figure 5.4.1-2** shows the hybrid model estimating the p-y curve for liquefied soil.

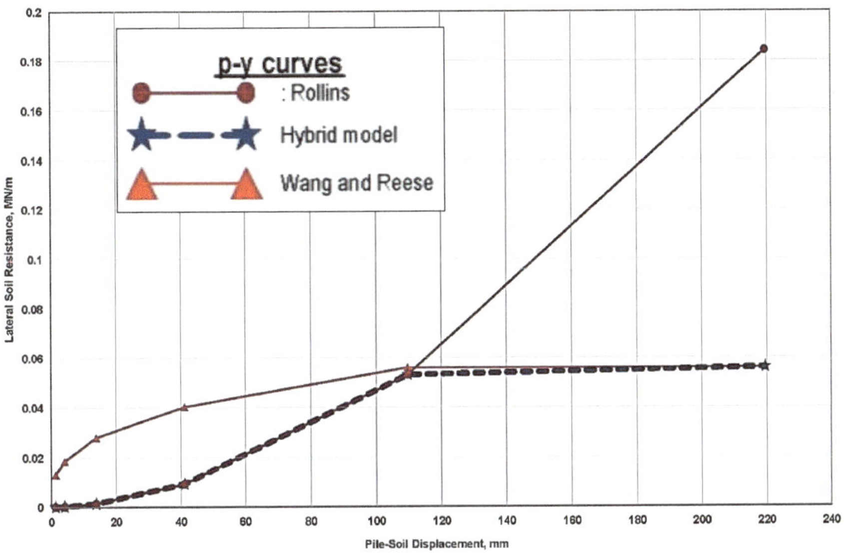

Figure 5.4.1-2: Hybrid P-Y for Liquefied Soils

Degraded load transfer curves (p-y, t-z, q-z)

Upon review of the geotechnical boreholes, the soil stratigraphy was primarily characterized as consisting of three units A, B, and C as shown in **Figure 5.4.1-3**. Unit A consists of Very soft to Soft Clayey Silt to Silty Clay in the top 15m. Unit B consists of Medium Dense silty Sand from 15m to about 30m and Unit C consists of alternating layers of very dense silty Sand/Stiff Clay.

In cohesive soils (Unit A), p_u calculation relies on undisturbed shear strength. However, seismic events can induce significant displacement at the mudline, disturbing soil-pile interaction in the near vicinity and reducing its strength to near remolded value. Hence, utilizing remolded shear strength for pu estimation (approx. from 0-15m) is advisable, which is derived from lab vane tests.

For the medium-dense sand layer (Unit B), although the relative density is high, suggesting no liquefaction hazards, there are weak layers (derived from CPT) within Unit B having a relative density of less than 50%. Therefore, Unit B was considered to undergo partial liquefaction. Thus, the hybrid p-y model was used to estimate the degraded P-Y curves.

Below 30m (Unit C), no degradation was considered since soil stratigraphy is strong enough not to have any susceptibility to liquefaction.

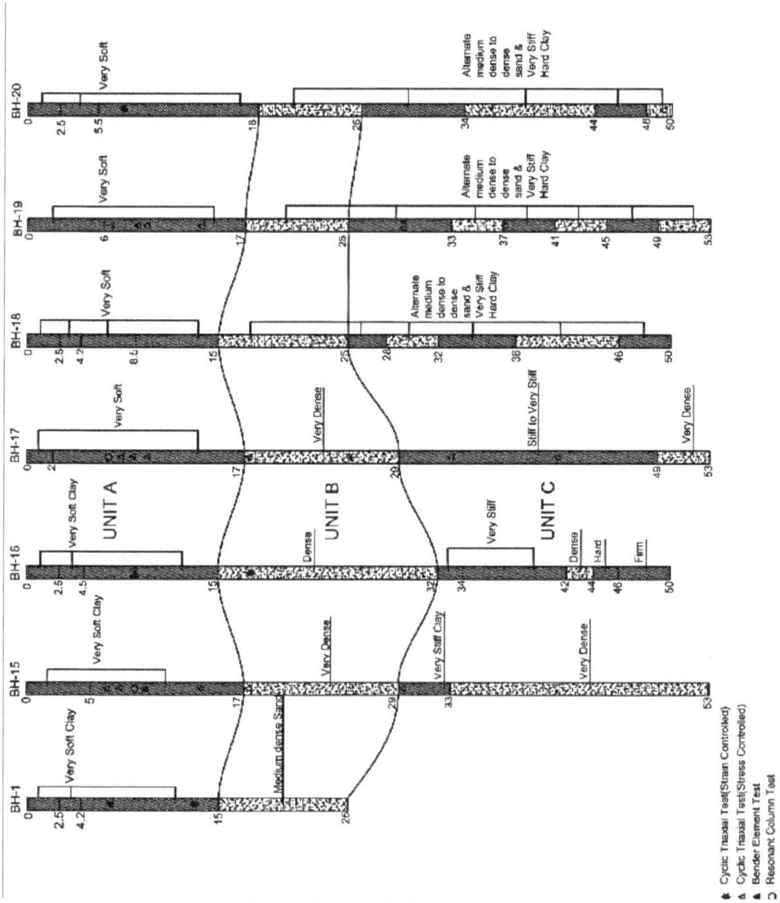

Figure 5.4.1 3: Soil Stratigraphy

Like p-y, the degraded models for predicting t-z and q-z behavior are not available. Therefore, in the absence of such models, it seems reasonable that the degraded models (similar to p-y) be assumed to characterize the effects of liquefaction on t-z and q-z behavior as well.

The design parameters used to develop degraded p-y, t-z, and q-z data are shown in **Table 5.4.1-2** Since Medium Dense Sand layers exist beyond 15m, the residual shear strength method prevails in the use of hybrid model, as Rollins method provides higher values **Figure 5.4.1-2** above shows the sample degraded p-y curves for one borehole (Hybrid curve).

Depth	Soil type	qc, MPa	D_{50}, mm	qc/ (Pa*N)	Correlated N Value, bl/0.3m	Su, kPa	γ', kN/ m³	φ	Remoulded shear strength, kpa	Residual Shear strength , kPa	p-multiplier	Relative Density
0.0-2.5	Clay	0.2	0.028	1.5	1	4	6.5	-	2-3	-	-	-
2.5-4.2	Clay	0.2	0.028	1.5	1	7	6.5	-	4	-	-	-
4.2-15	Clay	0.2	0.038	1.8	1	12	5.5	-	4	-	-	-
15.0-17.0	Sand	5	1	8	6	-	7.0	30	-	4.78	-	45.27
17.0-19.0	Sand	40	5	8	50	-	7.0	30	-	-	0.5	128.04
19.0-25.0	Sand	15	5	8	19	-	7.0	30	-	45.89	-	78.41

Table 5.4.1-2: Soil design parameters for degraded p-y data

Figures 5.4.1-3, 5.1.4-4, 5.1.4-5 and 5.4.1-6 depicts lateral load-deformation curves (p-y) at different depths (5.0m, 13.0m, 17.0m and 20.5m respectively).

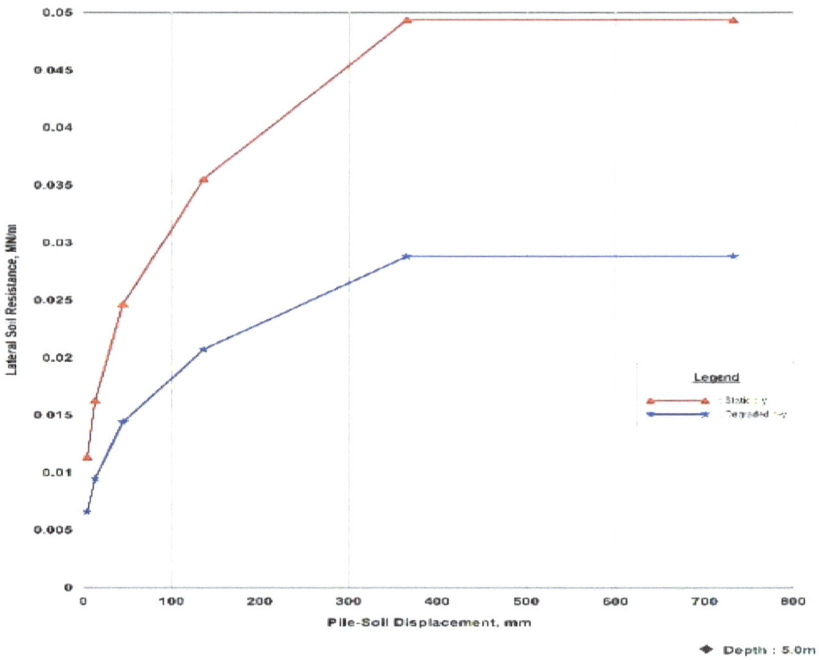

Figure 5.4.1-3: Lateral Soil Resistance Vs Displacement (5.0m)

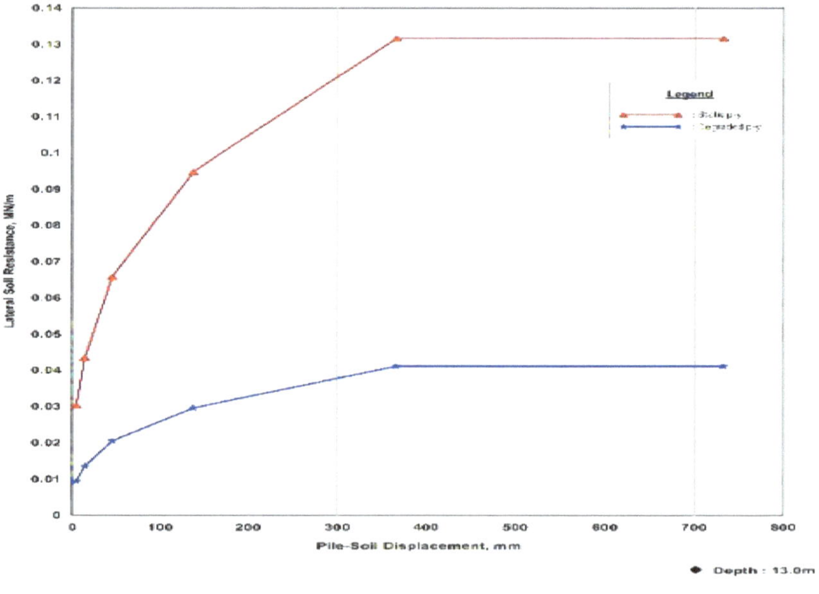

Figure 5.4.1-4: Lateral Soil Resistance Vs Displacement (13.0m)

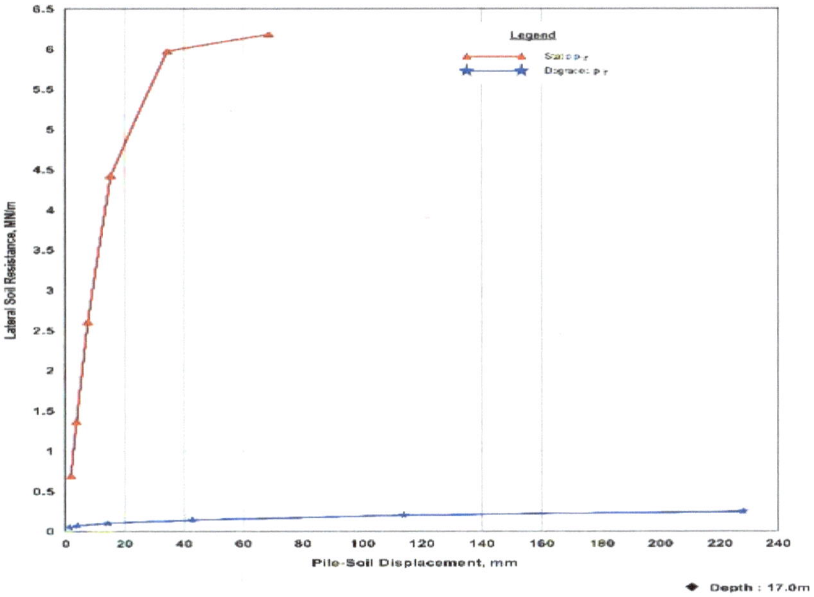

Figure 5.4.1-5: Lateral Soil Resistance Vs Displacement (17.0m)

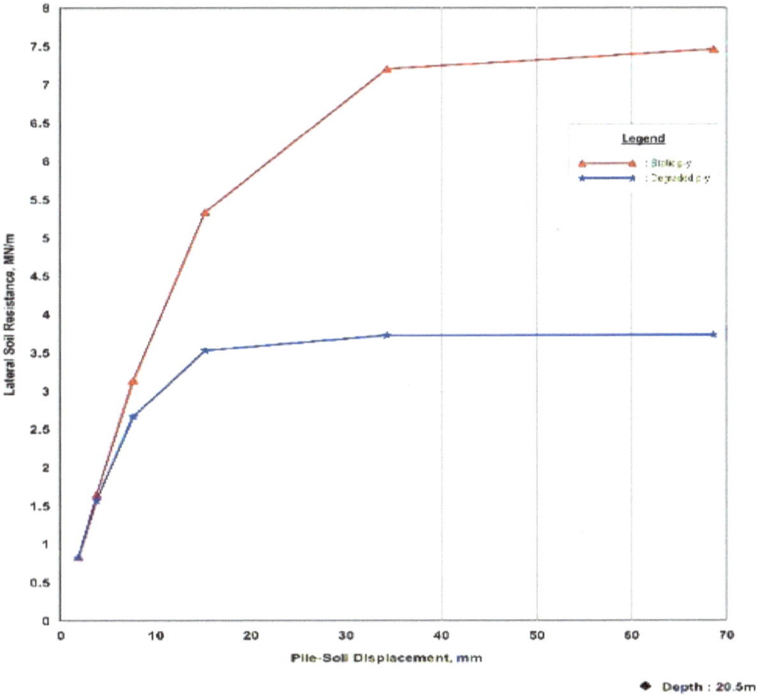

Figure 5.4.1-6: Lateral Soil Resistance Vs Displacement (20.5m)

Note: Although this case study is not pertaining to offshore Abu Dhabi, this is brought out to illustrate the method of degrading the p-y data due to seismic activity. The procedures outlined can be used even for offshore Abu Dhabi soils.

Conclusion

The degraded p-y curves are generally estimated using the hybrid model. However, for the present study, medium dense layers exist at beyond 15m. Therefore, the residual shear strength method prevails in the use of hybrid model, as Rollins method provided higher values.

Finally, anchor piles were successfully installed and performed very well.

5.4.2 Case Study – 2 Evaluation of As-Installed Pile Capacity of a Drilled and Grouted Pile installed short of expected termination depth. (Dr CR Parthasarathy, Ms Ashitha, Mr Sumanth and Mr Nandagopal)

Problem Statement

Assessment of the as-installed pile capacity is required when the pile is not installed to the design penetration due to encountering unexpected installation problems such as malfunctioning of installation equipment or complexity in predicting the site response.

This case study deals with the assessment of as-installed capacity for an offshore drilled and grouted pile that could not be installed to design penetration due to complex site conditions. This study demonstrates how the installation data such as pile driving parameters -blow counts, energy measurement from impact hammer console- and caliper data can be used in assessing the as-installed capacity. The hindcast analysis using GRL WEAP (Wave Equation Analysis for Piles) and the qualitative assessment of available grout thickness aided in evaluating the as-installed capacity and thus the available safety factors.

Drilled and Grouted Pile Installation Procedure and Capacity Estimation

Drilled and grouted piles have a great potential for utilization in carbonate sandy soils since such deposits are typically not suited for driven piles due to the very low shaft friction that results from the driving process (David et al., 2014).

These piles are suitable in cases where the piles can't be installed using an impact hammer if the soil profile consists of cemented soils, calcarenite, or gypsum layers.

It is usual to have a sleeve pile installed first using an impact hammer up to a certain depth to retain the loose soils present at shallow penetrations. Then this is followed by the drill-out of soil/rock up to design penetration, insert a pile having a smaller diameter than sleeve pile and grout the pile & the annulus between the pile & the soil/rock.

The general design method selected to calculate the axial pile capacity of Drilled & Grouted (D&G) pile is as per **API RP2GEO (R2021)**, Nauroy et al., 1985, Stevens 1996, Angemeer et al, 1973, Abbs, 1983 and Abbs & Needham 1985. The end bearing component of the drilled and grouted pile is not considered due to the inability to ensure that the pile bore base is sufficiently clean of cuttings and debris to enable full resistance to develop without excessive pile settlement.

The pile capacity of a drilled and grouted pile is estimated as the sum of the sleeve pile capacity (from 0m to final penetration of sleeve pile i.e., the driven pile capacity) and the insert pile capacity (for the depths between final penetrations of sleeve pile and insert pile i.e., the drilled and grouted pile capacity).

Project Details

The current study is from an offshore project site in the Middle East. The site-specific geotech borehole indicates that stratigraphy predominantly consisted of Calcarenite, Calcisilitite, and Calcilutite.

The soil parameters considered for the design of drilled and grouted pile are shown in **Table 5.4.2-1.**

Layer	Soil/ Rock Type	Depth (m)		Submerged Unit Weight (kN/m3)	Shear Strength (kPa)		CaCO3	Soil-Pile Friction Angle
		From	To		From (Top)	To (Base)	(%)	
1	Sand	0	5	8.3	-	-	100	12.5
2	Sand	5	6.7	9	-	-	100	30
3	Calcarenite	6.7	8	9.9	-	-	93	25
4	Calcisilitite	8	8.55	11	-	-	100	25
5	Clay	8.55	10	8.9	250	250	-	-
6	Sand	10	11	8.9	-	-	93	20
7	Silt	11	14	10.4	500	500	-	-
8	Calcisilitite	14	14.92	10.6	-	-	78	25
9	Calcarenite	14.92	17.5	9.4	-	-	93	25
10	Calcisilitite	17.5	19.5	10.4	-	-	100	25
11	Clay	19.5	30.5	9.6	160	160	-	-
12	Clay	30.5	40.3	9.7	160	200	-	-
13	Sand	40.3	41.8	8.9	-	-	34	35
14	Clay	41.8	42.6	9.5	300	300	-	-
15	Calcilutite	42.6	43.5	10.8	-	-	72	25
16	Calcisiltite	43.5	46.3	8.6	-	-	100	25
17	Silt	46.3	48.15	11.1	500	500	-	-
18	Calcarenite	48.15	48.5	11.1	-	-	93	25
19	Calcisiltite	48.5	55	10.8	-	-	78	25
20	Silt	55	56	10.8	500	500	-	-
21	Sand	56	57	10.8	-	-	80	22.5
22	Silt	57	58	10.8	500	500	-	-
23	Calcisiltite	58	59.5	8.4	-	-	56	25
24	Silt	59.5	60	8.6	400	400	-	-
25	Calcisiltite	60	61.3	8.6	-	-	72	25
26	Silt	61.3	64	8.6	150	150	-	-
27	Calcisiltite	64	68	10.5	-	-	93	25
28	Silt	68	72.06	8.9	300	300	-	-

Table 5.4.2-1: Geotechnical Design Soil Parameters

A drilled and grouted pile was designed to consist of a sleeve pile of 1.219m diameter and 25mm wall thickness and an insert pile of 1.067m diameter and 25mm wall thickness to be placed into a grout hole of diameter 1.217m. The sleeve pile was first driven to a target penetration of 14.5m below the mud line. A 1.2m diameter hole was drilled through the sleeve pile up to the design penetration of 57.95m below the mud line.

A caliper measurement of the grout hole was carried out inside the drilled hole to confirm the drilled shaft diameter along the entire depth of the hole. The caliper data suggested no major anomalies during the lowering and retrieving of the Koden apparatus inside the drilled hole and the average diameter available was 1.2m. However, the drilled hole had a slight eccentricity ranging between 19 to 40mm, and the deviation from the vertical is less than 1%. After the caliper measurement, the main pile of 1.066m outer diameter and 25mm wall thickness was inserted through the sleeve pile and the pile rested at 57.95m.

However, the pile was lifted to reorient the weld seam as per the construction requirements. The pile was inserted back into the drilled hole but faced difficulties reaching the target depth of 57.95m. The pile could only go up to 54.80m depth whereas the design depth was 57.95m thus a pile length reduction of 3.15m occurred.

Even the additional measures such as high-pressure flushing of the bottom with 30MT of water and placing the weight of 46MT (S-280 hammer) on the top could not advance the insert pile further. No cuttings were observed during the flushing indicating that the borehole did not collapse. The pile bottom was stuck in the Calcilutite, which has a UCS value of 3MPa, as mentioned in the geotechnical report.

The insert pile's lifting, re-alignment, and lowering operations were suspected to have caused the dislocation of jointed rock mass resulting in the obstruction. With the pile resting at 54.80m, the grouting was carried out and the grout volume consumed was about 72 % more than theoretical grout consumption. As the insert pile did not reach the design penetration, concerns were raised about the adequacy of the as-installed pile capacity to meet the minimum required safety factors.

This warranted the root cause analysis for the borehole obstruction, the assessment of the available grout thickness, and the as-installed capacity of the drilled and grouted pile using the available information viz., driving records of the sleeve pile, caliper measurements of the drilled hole, grout volume consumed, and insert pile depth.

Review of Available Information

Upon Review of records and available information, it was understood that the required pile capacity at 54.8m pile penetration is 53.5 MN. The caliper logs from both during lowering and retrieving did not show major anomalies except for a slight eccentricity of about 19 to 40mm (with different trials of Caliper logs) and verticality of <1%. The eccentricity and verticality mentioned are of the drilled hole but not of the insert pile itself. The average hole diameter is about 1.2m (48") as shown in **Table 5.4.2-2.** This confirms that the socket was in line with the drawing requirement.

Particulars	Value	Remarks
Average Grout Hole Dia (m), (Measured from caliper logs)	1.2	-
As installed Pile length below mud line (m)	54.8	-
Eccentricity (mm)	40	w.r.to Drilled Hole Centre at the top (Negligible)
Verticality (%)	0.072	w.r.to Drilled Hole Centre at the top (Negligible)
% Variance in diameter as per caliper details	1.93	w.r.to Dilled Hole Diameter (Negligible)

Table 5.4.2-2: Caliper Logging Details

Grouting records showed a 72% increase in the actual cement consumption (35 tons) compared to the computed cement content (20.37 tons), and a 64% increase in the actual water pumped out relative to the calculated quantity of water. This confirms that the borehole was filled with grout beyond 54.8m. The average grout thickness along the length of the pile is also estimated to be about 75mm except for about 4.8m in 1/4th of the pile circumference towards the bottom as shown in **Figure 5.4.2-1** which may have a minimum thickness of the grout.

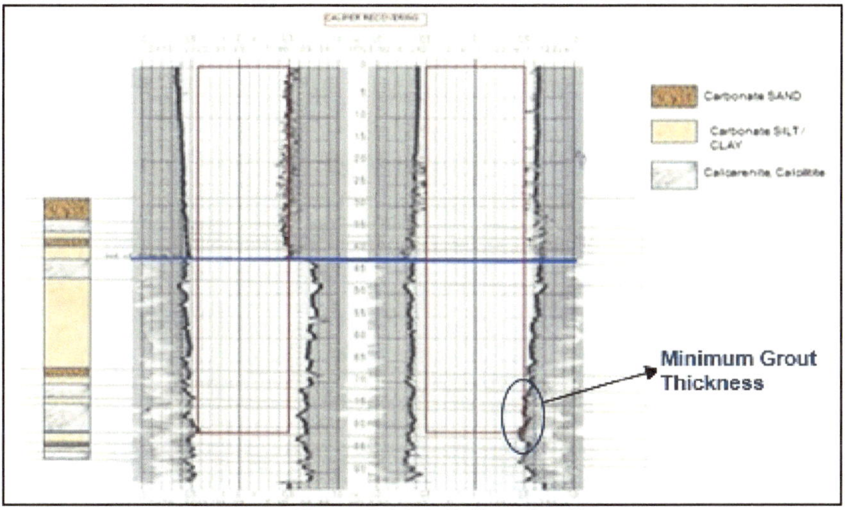

Figure 5.4.2-1: Caliper details

Root Cause Analysis

The possible reasons for the pile not penetrating to design depth are illustrated below **Table 5.4.2-3.**

The possible reason for the pile not penetrating to design depth	Root cause	Yes/No
Insufficient borehole depth	shorter depth of borehole drilled Incorrect depth measurement Incorrect recording	No No No
Borehole collapse	Borehole Collapse Improper consistency of the drilled fluid Issues with the pump Insufficient mud Loose soil layers **Dislocation of jointed rock (CALCISiltITE)**	No No No No No Yes
Improper grout hole construction	Wrong diameter **Eccentricity** Absence of Caliper Measurement	No **within Limits** No
Pile Buckling	Improper fabrication Poor material quality Wrong diameter Driving pile in a hard layer	No No No No
Verticality issues of pile	Eccentric borehole Pile is bend	No No

Table 5.4.2-3: Root Cause WHY analysis

From the above root cause analysis, it is evident that the dislocation of jointed rock could be the only possible reason for impeding the pile penetration below 54.8m.

Assessment of Drilled and Grouted Pile Capacity

The as-installed drilled and grouted pile capacity is computed as the sum of the driven sleeve pile capacity and the grouted insert pile capacity.

Assessment of Sleeve Pile Capacity

The pile driving record indicates that the sleeve pile penetrated to 5.25m due to self-weight. Then, an IHC S-280 Impact hammer was used to drive the pile up to 14.2m penetration at hammer efficiencies ranging from 15% to 35%. Since the pile was not instrumented to measure the force and velocity signature for each hammer blow to estimate the available pile capacity, a hindcast analysis was performed.

Hindcasting is the simulation of the soil behavior that occurs during pile driving using wave equation software (GRL WEAP). The soil model is adjusted until the blow counts determined are matched with the actual blow count for the measured hammer energy. The end-of-drive blow counts from 14.0m to 14.2m penetration was 8 blows/0.2m. For 2.8MN skin friction capacity, hammer energy is 93.14KJ which is 33% efficiency. The blow count for 13.95m to 14.20m penetration is 10.9 blows/0.25m. From linear interpolation, computed blows for 0.20m penetration is 8.8blows/0.2m, which matches the field pile driving record. Hence, the blow count matching technique confirms that the driven pile capacity is about 2.8MN **Figure 5.4.2-2** and **Figure 5.4.2-3.**

Ultimate Capacity kN	Maximum Compression Stress MPa	Maximum Tension Stress MPa	Blow Count blows/.25m	Stroke m	Energy kN-m
2800.0	115.03	23.36	10.9	2.08	93.14

Figure 5.4.2-2: Results of Hindcast analysis using GRL WEAP

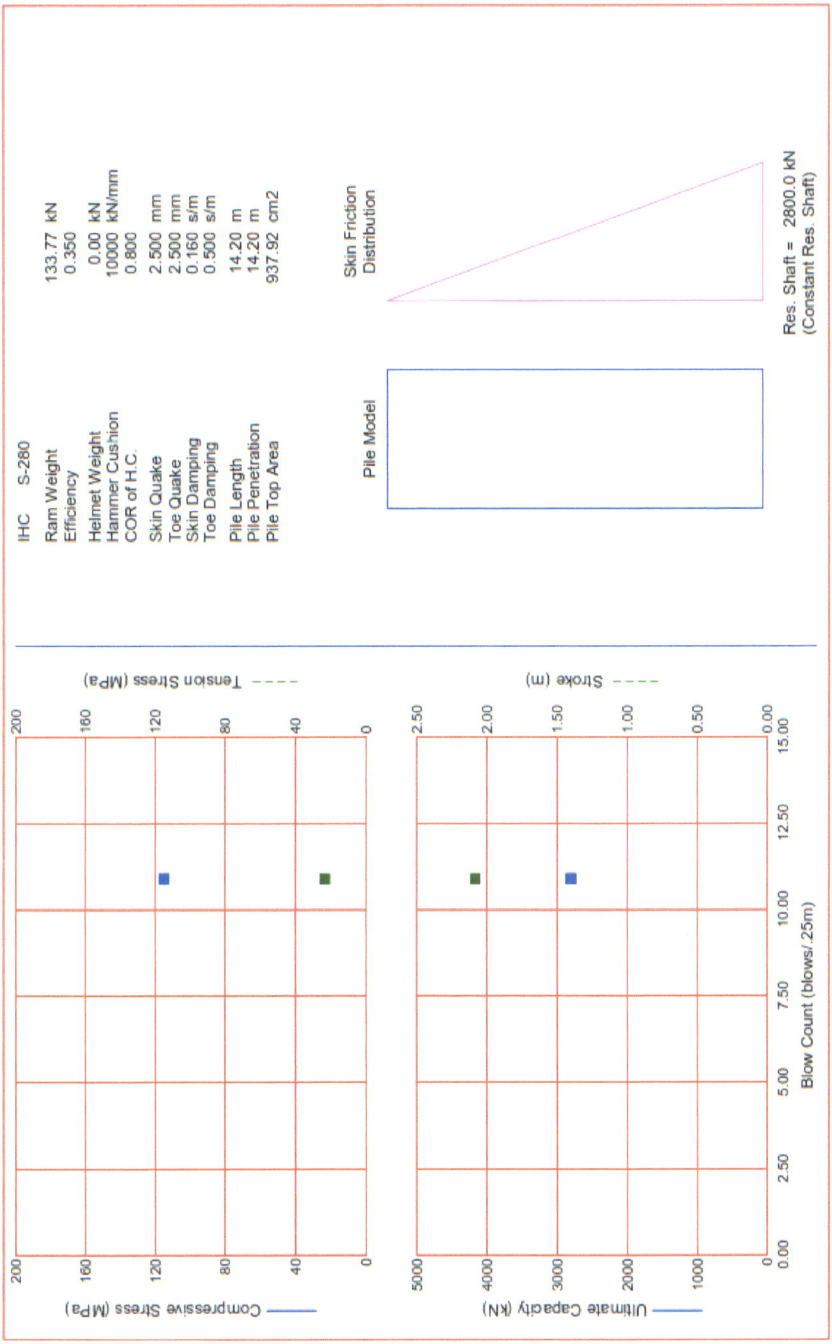

Figure 5.4.2-3: Results from GRLWEAP

Assessment Of Pile Capacity of Drilled and Grouted Pile

In addition to the considerations presented for granular and cohesive soils in **API RP2GEO (R2021)**, shaft friction in rocks & limiting values of the bond between the steel insert pile and the grout, the grout & the surrounding rock/soil are computed as presented in **Figure 5.4.2-4.**

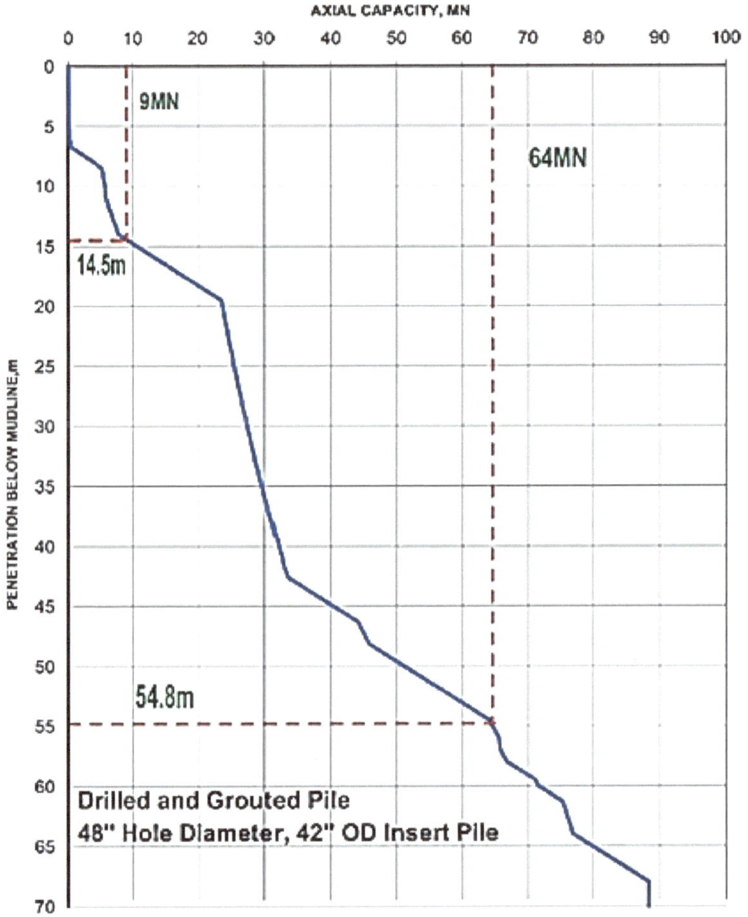

Figure 5.4.2-4: Axial Capacity vs penetration of Drilled & Grouted Pile (Leg A1)

The as-installed D&G pile capacity is calculated as shown below **Table 5.4.2-4** Saudi Aramco Standards SAES Q-014 was also referenced as standard guidelines for this present study.

Description	**Capacity (MN)**	**-**
Main Pile Capacity- Drilled and Grouted Pile Capacity @ 54.8m	64	(A)
Main Pile capacity- Drilled and Grouted Pile Capacity @ 14.5m	9	(B)
Sleeve Pile resistance- Driven Pile Capacity @14.5m	2.8	(C)
Drilled & Grouted Pile Capacity @ 54.8m	57.8	((A)-(B)) +(C)
Drilled & Grouted Pile Capacity Without Sleeve Friction @ 54.8m	**55**	**((A)-(B))**
C is neglected for a conservative estimate of pile capacity		

Table 5.4.2-4: Drilled and Grouted Pile capacity

Qualitative Assessment Considering 75% D&G Pile Capacity and 25% Driven Pile Capacity for the Bottom 4.8m Pile Length (With and Without Sleeve Pile Capacity).

The sensitivity assessment of pile capacity was conducted by considering 75 % of D&G pile capacity and 25% of Driven pile capacity for the bottom 4.8m pile length (due to marginal thickness in 1/4th of the pile circumference as per caliper and installation details as shown in **Table 5.4.2-4** Total pile capacity without considering the sleeve pile, was estimated to be 51.84MN with a FOS of 1.96. However, with the addition of a sleeve pile capacity of 2.8 MN, a FOS of 2.07 is established **Table 5.4.2-5.**

A	Driven	**Sleeve pile capacity in top 14.5m**	2.8	MN
B	D& G pile	Pile capacity at 50.0m	50.74	MN
		Pile capacity at 14.5m	9	MN
		D&G pile capacity for 35.5m (50.0-14.5)	41.74	MN
C	D& G pile	D& G pile capacity (54.8)	64	MN
		Pile capacity at 50.0m	50.74	MN
		D& G pile capacity for 4.8m (54.8-50)	13.26	MN
		75% of D&G pile capacity for 4.8m	**9.95**	MN
	Driven pile	Driven pile capacity (54.8)	13.39	MN
		Pile capacity at 50.0m	12.69	MN
		Driven pile capacity for 4.8m (54.8m-50)	0.61	MN
		25% of Driven pile capacity for 4.8m	**0.15**	MN
		75% D&G pile + 25% Driven pile	10.1	**MN**
FOS (Without Sleeve Pile)	1.96	**Total capacity (B+C)**	51.84	**MN**
FOS (With Sleeve Pile)	2.07	**Total capacity (A+B+C)**	54.64	**MN**

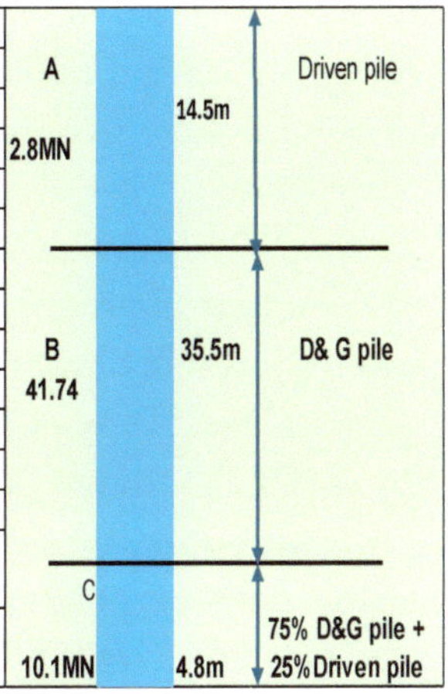

Figure 5.4.2-5: Driven and D&G Pile capacity

Summary and Conclusions

The pile under examination concluded its penetration at 54.8m, deviating from the initially planned 57.95m. The caliper logs, recorded during both lowering and retrieving operations, revealed no major anomalies, except for a slight eccentricity ranging from 19mm to 40mm (varied in different caliper log trials) and verticality of less than 1%, which falls within acceptable limits in the normal deep foundation industry. It's crucial to note that the mentioned eccentricity and verticality refer to the drilled hole itself, not the pile. The average hole diameter, measuring about 1.2m (48"), confirms that the socket aligns with the specified drawing requirements.

Grouting records indicate a substantial 72% relative increase in the actual quantity of cement consumed (35 tons) compared to the computed cement content (20.37 tons). Similarly, there is a 64% relative increase in the actual quantity of water pumped compared to the computed quantity. The higher actual grout consumption, exceeding 70% relative to theoretical consumption, signifies effective grout penetration to the pile base and the socket length. The estimated capacity of the driven sleeve pile at a 14.5m penetration is approximately 2.8MN (analytical). Hindcast analysis, aligning blow counts with the energy from the hammer console using GRL WEAP software, closely agrees with the pile driving record for the analytically estimated 2.8MN skin friction capacity. Considering the as-installed pile penetration of 54.8m for a 48" diameter hole, grouted with an insert pile of 42" diameter, results in an estimated Drilled & Grouted pile capacity of about 55MN below the sea floor, exceeding the required pile capacity of 53.5MN. A sensitivity assessment, considering 75% of the Drilled & Grouted pile capacity and 25% of the Driven pile capacity for the bottom 4.8m pile length, results in an estimated total pile capacity of approximately 51.84MN with a FOS of 1.96. However, with the sleeve pile, a higher FOS of 2.07 is established.

Given that the requisite Factor of Safety is met for an installed pile penetration of 54.8m, the pile was finally considered acceptable.

5.4.3 Case Study – 3 - Pipe-soil interaction study (Dr Shashank Mouli, Dr Prashanth Talkad & Dr CR Parthasarathy)

Introduction

This field of project represents a significant offshore oil and gas reservoir critical to UAE's energy sector. This development project is an integral part of the overall efforts to develop these untapped hydrocarbon reserves in UAE, with priority given to produce 340 MMscfd of raw gas by 2024.

Problem Statement

Offshore on-bottom pipeline-soil interaction study is done to understand the breakout resistance, axial and lateral soil resistances to pipe embedment and movements. It is necessary to study the available geotechnical and geophysical investigations performed along the proposed pipeline routes and derive the most appropriate geotechnical parameters specific to the pipeline chainages (zoning of soils based on seabed properties). It is a challenge to arrive at the ground model if the seabed soils consist of Calcarenite or cemented sands. In such cases, it is appropriate to study the nearby on-bottom pipeline penetrations, if available and modify the geotechnical parameters accordingly. This case study discusses the methodology of arriving at the ground models, determination of pipeline embedment, equivalent axial and lateral soil friction coefficients, static and dynamic seabed stiffness for the proposed 52.406km stretch of subsea pipeline in offshore Abu Dhabi.

Methodology

The assessment of geotechnical parameters, Pipe-Soil Interaction forces for submarine pipelines through PSI studies form integral part of the pipeline design. The parameters are derived from site specific geotechnical field investigations and laboratory test results. The DNVGL has published recommended practice **DNVGL-RP-F114, 2019** titled "Pipe Soil Interaction for Submarine Pipelines" providing basic guideline on the acceptable methodologies to derive the soil pipeline interaction parameters.

This case study presents the results of the Monotonic cycles of PSI evaluated for a stretch of subsea pipelines along with the methodology and is called Pipe to Soil Interaction. The geotechnical results indicate that the soils up to two times the diameter of the Pipe are cohesionless soils (Calcareous Sand/Carbonate Sand/Calcarenite) and hence the calculations are based on drained loading condition only and it is not a time dependent phenomenon.

Pipeline Embedment

The following scenarios that can happen during the pipeline installations are considered for their penetration assessment into the seabed.

- The static self-weight burial of the pipeline.
- Load concentration caused by the dynamic impact of the pipeline at the time of installation.
- Additional dynamic effects such as lay barge induced vertical and horizontal movements if any.

The presence of additional vertical loads and dynamic lay effects during the installation process may cause an increase in pipe embedment depth. DNV-RP-F114 suggests that such external factors may contribute to increase in pipe penetration by a touch down lay factor varying between 1 and 3.

Drained Loading Condition

The drained loading condition is applicable for cohesionless soils. The static self-weight burial of the pipeline and bearing capacity components are estimated as per DNVGL–RP-F114, 2019, relating the vertical force, Qv required to penetrate the pipe to an unknown embedment, z.

An external load factor is also considered in the estimation of pipeline embedment by adopting a touchdown factor (k_{lay}) as Qv/Wi; that relates vertical force with seabed stiffness, lay tension, departure angle and pipeline bending stiffness in addition to the static force to the weight of the pipe (Wi).

Estimation of k_{lay} factor

Touchdown factor (K_{lay}) is basically a multiplication factor to the static vertical force to account for the additional force that the pipe will impose on to the soil during the touchdown of seabed in the installation process. This will be estimated as per the equation specified in DNVGL-RP-F114 - Section 4.2.5.2.

It is assumed that there is no embedment effect as long as the penetrated pipe is inside the active Rankine zone. The depth effect is usually not important in sands, because of the large penetration needed before it is relevant (DNVGL–RP-F114, 2019).

Additional embedment due to dynamic laying effect (f_{dyn} or k_{lay} dynamic)

In case of drained loading, the additional embedment due to dynamic laying effect is characterized by the combined vertical and lateral motions of pipeline occurring at the touchdown zone during installation. This is directly related to the sea state during laying (DNVGL-RP-F114). As such, if the pipeline is laid during the rough sea state the embedment is likely to be more than the pipeline laid during the calm sea state. The additional embedment shall be defined using a multiplicative dynamic embedment factor, f_{dyn}. (Westgate et al., 2012). Values reported in the literature generally range between 1 and 8 for fine-grained soils based on as-laid surveys. For fully drained carbonate sands, f_{dyn} typically ranges between 3 and 8 (Westgate et al., 2012). The additional embedment due to the dynamic laying effect is considered by multiplying f_{dyn} values with $w_{calculated}$. Accordingly, the Low (LE), Best (BE) and High (HE) estimate of additional embedment depth will be estimated. Considering the loose to medium dense nature of the ground along the route, f_{dyn} is taken as 3, 4, 5 for LE, BE and HE respectively. For the stretches with Calcarenite as ground model, the f_{dyn} is taken as 1 (one) as the embedment is negligible. The dynamic embedment factor, f_{dyn} for the CSI (cable-soil interaction) analysis is taken as 1 (one). If the weight of the pipe during the installation/empty condition including lay effects is more than the weight of the pipe during the flooding condition, then the embedment depths that are calculated from the installation/empty condition shall be used for the flooded/ operation condition also. It is worth noting that the effect of dynamic amplification factor on spools is quite marginal to cover soil and pipeline interaction. Moreover, all the uncertainties in the soil behavior are accounted by considering Low, Best and High estimate parameters. Higher spool embedment due to dynamic factors indicates more passive resistance which consequently results in tighter curvatures on spools (i.e. higher stresses) and more resistance to pipeline expansion (i.e. higher pipeline effective force). Hence, no separate calculations were provided for spool.

Equivalent Axial Friction Coefficient between Pipe and Soil

The axial friction coefficient is defined as the ratio between axial resistances to the vertical pipe soil force. The drained loading condition is applicable for cohesionless soils while the undrained loading condition is applicable for cohesive soils. Theoretically, drained and undrained responses are idealized with bilinear and trilinear force displacement curves. Pipes resting over dense sand or stiff clay will have distinct breakout and residual resistances whereas for pipes resting on loose sand or soft clay, these resistances generally converge at the breakout displacement. The total pipe - soil contact force can increase up to 1.27 times due to curved surface of the pipe; and is accounted by introducing a wedging factor, ζ (White and Randolph, 2007). Lower and upper estimates of the axial friction coefficient are estimated by defining the lower and upper bound ground model from the range of soil parameters obtained from field and laboratory geotechnical data for the intended pipeline route.

A) Drained Loading Condition

The drained axial friction coefficient is calculated as per DNVGL–RP-F114, 2019, concrete pipe roughness factor ($r_{pipe\text{-}soil}$) of 0.8, 0.9 and 1.0 are considered as low, best and high estimates respectively for PSI calculations for the intended pipeline route. The residual friction coefficient ($\mu_{A,res,d}$) is considered by using residual friction angle instead of peak friction angle for the estimation as suggested in **DNVGL–RP-F114, 2019.**

B) Axial Mobilization Displacements

In structural modelling of pipeline, the axial response is modelled using bilinear or trilinear elastic perfectly plastic model. In such cases, it is necessary to define the mobilized displacement. For the intended pipeline route PSI analysis, the axial breakout and residual resistances are mobilized at distances **(DNVGL–RP-F114, 2019)** as follows:

The axial breakout & axial residual displacements are calculated based on peak friction angles. As per the DNVL-RP-F114, 2019, Bi-linear soil is to be considered for drained condition of loading. **Table 5.4.3-1** present the axial mobilization displacement for Bi-linear soil model respectively.

Axial model	**Uncertainty case**	**Parameter**	**Typical values**
Bi-linear (i.e. with no breakout peak)	Low estimate, LE	X_{mob}	Minimum of 1.25 mm and 0.0025D
	Best estimate, BE		Minimum of 5 mm and 0.01D
	High estimate, HE		Maximum of 250 mm and 0.5D

Table 5.4.3-1: Axial Mobilization Displacements for Bi-linear Soil Model

Equivalent Lateral Friction Coefficient between Pipe and Soil

The lateral response of the pipe is characterized in two stages.

- First load displacement (monotonic) is characterized by breakout resistance and a steady residual resistance.
- Cyclic displacement is characterized by the growth of soil berms at the limits of the pipe displacement range, with the pipe descending into a shallow trench. The loading scenarios may be valid for repeated heating and cooling corresponding to breaks in production at the location of the pipe where lateral buckle displacement occurs.

Lower and upper estimates of the lateral friction coefficients are estimated by defining the lower and upper bound ground model from the range of soil parameters obtained from field and laboratory geotechnical data for this pipeline route. Since the ground model for the entire pipeline route is Sand/ Calcarenite, only drained loading conditions are considered for evaluating the lateral breakout/residual friction coefficient.

Lateral breakout friction coefficient:

The drained lateral breakout friction coefficient between pipe and soil ($\mu_{L,brk,d}$) is defined as the ratio of drained lateral breakout resistance of the soil to the vertical pipe soil force.

$$\mu L,brk,d = \frac{F_{L,brk,d}}{V} \qquad \textit{Equation 5.4.3-1}$$

Where,

FL,brk,d = Lateral Breakout resistance

V = Vertical force of Pipe - soil for the condition analyzed

The drained lateral breakout resistance is the summation of the active resistance, i.e. the frictional resistance between pipe and soil; and the passive resistance. The lateral breakout resistance (FL,brk,d) is estimated using Model 2 given in DNVGL–RP-F114, 2019. However, the frictional term 0.6V is replaced with $r_{pipe-soil} \times tan\varphi_{peak} \times V$ since the pipes are coated with concrete. Pipe-soil roughness factors (rpipe-soil) of 0.8, 0.9 and 1.0 are considered as low, best and high estimates respectively for PSI calculations.

Lateral residual friction coefficient:

The drained lateral residual friction coefficient between pipe and soil ($\mu_{L,res,d}$) is defined as the ratio of drained lateral residual resistance of the soil to the vertical pipe -soil force.

$$\mu L, res, d = \frac{F_{L,resd}}{V} \qquad \textit{Equation 5.4.3-2}$$

Where,

FL,res,d= Lateral residual resistance

V = Vertical force of Pipe - soil for the condition analyzed

The drained lateral residual resistance is the summation of active resistance, i.e. the frictional resistance between pipe and soil, and passive resistance. The lateral residual resistance (FL,res,d) is estimated using the procedure of lateral breakout resistance (Model-2 given in DNVGL-RP-F114). However, the frictional term 0.6V is replaced with $r_{pipe-soil}\ tan\varphi_{res}V$ since the pipes are coated with concrete. Pipe-soil roughness factor ($r_{pipe-soil}$) of 0.8, 0.9 and 1.0 are considered as low, best and high estimates respectively for PSI calculations.

As per DNVGL-RP-F114, zero embedment can be assumed in passive resistance calculations which will be over conservative for the analysis. Hence 50% of embedment depth is considered in passive resistance calculations in this study.

Lateral Mobilization Displacements

As per (DNVGL-RP-F114,2019), the lateral mobilization displacement is estimated as in **Table 5.4.3-2** considering various uncertainty cases.

Uncertainty case	Parameter	Typical values
Low estimate, LE	Y_{brk}	D(0.004 + 0.02z/D)
Best estimate, BE		D(0.02 + 0.25z/D)
High estimate, HE		D(0.1 + 0.7z/D)
Low estimate, LE	Y_{res}	0.6D
Best estimate, BE		1.5D
High estimate, HE		2.8D

Table 5.4.3-2: Lateral Mobilization Displacements

Shear Modulus

The shear Modulus (G) is calculated by using G_{max} at 1% of strain level. At 1% of strain level the G/G_{max} is 0.1. The G_{max} is calculated as per DNVGL-RP-F-114, 2019.

Static Soil Stiffness

Static vertical stiffness, KV,s is given as Qv/z; where Qv is the self-weight of the pipe and z is the embedment depth. As per DNVGL-RP114, 2019, static lateral stiffness with respect to breakout lateral resistance and residual lateral resistances are estimated.

Dynamic Soil Stiffness

The dynamic vertical and lateral soil stiffness are calculated as per DNVGL–RP-F114, 2019.

Shallow Foundation Methodology

Cohesionless Soils (Drained Behavior)

According to **API RP2 GEO(R2021**), the ultimate bearing capacity on Sand (Cohesionless Soil) is calculated considering the Load Inclination factors, Shape Factors, Depth factors, Base and seafloor surface inclination factors, Base inclination factors, and Sea floor shape factors.

Scour/Sediment Transport

The sediment transportation is anticipated when scouring is expected at the site. Scouring occurs if the velocity of the water particles at the seabed (V_{total}) is greater than the critical soil particle entrainment velocity (Uc) (Argema, 1993). However, in this project, $V_{total} < Uc$.

Hence no Scour is anticipated. Therefore, sediment transportation is also not anticipated.

Limitations on shallow foundation bearing capacity at pipeline crossings

The ultimate bearing capacity has been carried out assuming,

- For the pure vertical case, a uniform applied bearing stress and that all loads will act centrically upon the footing, i.e. no account for horizontal, inclined, or eccentrically loaded conditions has been taken.

- The seabed is flat and free from depressions or hollows which would prevent full contact being developed between the footing and seabed.
- No account has been made for unusual shapes that result in eccentric loading despite uniform loading.

Factors of safety were not included in the bearing capacity calculations and results.

Ground Model:

The generalized ground models for this pipeline route for the installation of 6" gas pipeline is derived based on field & laboratory data. The ground models were extracted from Factual Geotechnical Report. The entire route is discretized according to the pipeline outer diameter variation. The ground models are then summarized, and a typical ground model is shown in the below **Table 5.4.3-3.**

Borehole numbers	BH-06		
KP, m	From		To
(6" Gas Pipeline)	KP-00 (0.000)		KP-01 (670.00)
Soil type @ Mudline	MEDIUM DENSE Sand		
Average Water content, %	31.00		
Average Carbonate Content, %	56.00		
Particle size D_{50}, mm	0.31		
Specific gravity	2.64		
Elastic modulus, MPa	48.00		
Poisson's ratio	0.30		
	Low Estimate	**Best Estimate**	**High Estimate**
Bulk Unit weight, kN/m³	17.00	18.00	20.00
Submerged Unit weight, kN/m³	7.00	8.00	10.00
Angle of friction, deg	33	35	37
Residual friction angle, deg	33	35	35
Void ratio	0.50	0.40	0.35

Table 5.4.3-3: **Ground Model**

Results and Conclusions:

Based on the derived ground model, a typically estimated PSI parameters are presented in the **Table 5.4.3-4** below.

Sl.no	Soil Properties	Units	Low Estimate	Best Estimate	High Estimate
1	Submerged unit weight, γ'	kN/m³	7.00	8.00	10.00
2	Soil peak friction angle, ϕ_{peak}	deg	33.00	35.00	37.00
3	Soil residual friction angle, ϕ_{res}	deg	33.00	35.00	35.00
4	Poisson ratio, ν	-	0.30	0.30	0.30
5	r pipe-soil	-	0.80	0.90	1.00
6	Void ratio	-	0.50	0.40	0.35
	Pipe properties				
7	Total OD including coating	m	0.336	0.336	0.336
8	Elastic modulus x Moment of Inertia, EI	kN-m²	3915.13	3915.13	3915.13
9	Submerged empty weight of pipe	kg/m	162.15	162.15	162.15
10	Touchdown Tension, T_0	kN	294.18	294.18	294.18
	Output Parameters				
11	Bearing capacity factor, N_q	-	25.96	33.12	42.68
12	Bearing capacity factor, N_γ	-	35.00	47.75	65.78
13	Embedment due to submerged empty weight of the pipe during installation, Z_1	mm	9.97	6.32	3.64
14	$k_{lay1}=k_{lay2}$ as per DNV during installation	-	1.26	1.33	1.44
15	Vertical penetration resistance due to klay static ,Qv	kN/m	2.00	2.12	2.30

Sl.no	Soil Properties	Units	Low Estimate	Best Estimate	High Estimate
16	Embedment due to klay static as perDNV, Z_2	mm	12.62	8.49	5.28
17	Dynamic lay factor, f_{dyn}	-	3.00	4.00	5.00
18	Additional embedment due to dynamic effects, $Z_3 = Z_2 \times f_{dyn}$	mm	37.87	33.97	26.42
19	Final Embedment, Z=Max(Z1,Z2,Z3)	mm	37.87	33.97	26.42
20	Embedment angle based on Z, β	deg	39.27	37.11	32.60
21	wedging factor, ζ	-	1.08	1.07	1.05
22	Contact width based on Z ,B	m	0.21	0.20	0.18
23	Vertical pipe-soil force during installation condition, V	kN/m	1.59	1.59	1.59
24	Equivalent axial breakout coefficient; $\mu_{A,brk,d}$	-	0.56	0.67	0.79
25	Equivalent axial residual coefficient; $\mu_{A,res,d}$	-	0.56	0.67	0.74
26	Equivalent lateral breakoutcoeffient; $\mu_{L,br}$	-	0.68	0.79	0.90
27	Equaent lateraleidual coeffint; $\mu_{L,res,d}$	-	0.59	0.70	0.76
28	Axal Mob Disp Xmb/Xbrk-LE	mm	0.84	0.84	0.84
29	xial Mob Disp, Xmob/Xbrk-BE	mm	3.36	3.36	3.36
30	Axial Mob Disp, Xmob/Xbrk-HE	mm	250.00	250.00	250.00
31	Axial Mob Disp, Xmob/Xres-LE	mm	0.84	0.84	0.84

Sl.no	Soil Properties	Units	Low Estimate	Best Estimate	High Estimate
32	Axial Mob Disp, Xmob/Xres-BE	mm	3.36	3.36	3.36
33	Axial Mob Disp, Xmob/Xres-HE	mm	250.00	250.00	250.00
34	lateral Mob Disp, Ybrk-LE	mm	2.10	2.02	1.87
35	lateral Mob Disp, Ybrk-BE	mm	16.18	15.20	13.32
36	lateral Mob Disp, Ybrk-HE	mm	60.06	57.33	52.05
37	lateral Mob Disp, Yres-LE	mm	201.30	201.30	201.30
38	lateral Mob Disp, Yres-BE	mm	503.25	503.25	503.25
39	lateral Mob Disp, Yres-HE	mm	939.40	939.40	939.40
40	Vertical static stiffness of soil, $K_{,v,s}$	kN/mm	42.00	46.82	60.20
41	Lateral breakout resistance, $F_{L,brk,d}$	kN/m	1.08	1.26	1.43
42	Lateral residual resistance, $F_{L,res,d}$	kN/m	0.93	1.11	1.21
43	Lateral static stiffness of soil at breakout displacement, $K_{L,s}$; LE	kN/m/m	17.99	21.89	27.45
44	Lateral static stiffness of soil at breakout displacement, K, $_{L,s}$; BE	kN/m/m	66.78	82.56	107.28
45	Lateral static stiffness of soil at breakout displacement, K, $_{L,s}$; HE	kN/m/m	514.59	620.96	763.72
46	Lateral static stiffness of soil at residual displacement, $K_{L,s}$; LE	kN/m/m	0.99	1.18	1.29
47	Lateral static stiffness of soil at residual	kN/m/m	1.85	2.20	2.41

Sl.no	Soil Properties	Units	Low Estimate	Best Estimate	High Estimate
	displacement, $K_{L,s}$; BE				
48	Lateral static stiffness of soil at residual displacement, $K_{L,s}$; HE	kN/m/m	4.64	5.51	6.01
49	Mean effective stress	kPa	3.81	3.99	4.45
50	Shear Modulus, G	kN/m²	2566.95	3029.06	3419.10
51	Vertical dynamic stiffness of soil, $K_{V,d}$	kN/m²	3227.02	3807.96	4298.30
52	Lateral dynamic stiffness of soil, $K_{L,d}$	kN/m²	2536.14	2992.71	3378.07

Table 5.4.3-4: Typical PSI results of 6" Pipeline- 168.3mm OD-CWC-80mm - (00-670m)- Empty Condition

Bearing Capacity Assessment for Shallow Foundation:

The typical ultimate bearing capacities for different sizes and geometry (rectangle, L/B ratio = 2) for shallow foundation in this stretch is presented in the **Table 5.4.3-5 below and in Figure 5.4.3-1.**

	Ultimate Bearing Capacity*							
Existing Cable lne	30" Gas Pipeline – Across 42" Gas Pipeline – Across							
KP (km)	6" Export Condensate Pipeline – 33.800 6" Export Condensate Pipeline – 34.044							
Depth (m)	0.00							
Correction factor Kq for square	1.52							
Correction factor Kγ for square	0.60							
Correction factor Kq for rectangle	1.26							
Correction factor Kγ for rectangle	0.80							
Breadth of foundation (m) **	1.0	1.5	2.0	2.5	3.0	3.5	4.0	5.0
Qu, Rectangle (kN)	398	1343	3184	6219	10747	17065	25473	49753
Qu, Square (kN)	149	504	1194	2332	4030	6399	9552	18657

*Bearing capacities are ultimate unfactored.

** Breadth of foundation for Square is length of side

** Breadth of foundation for rectangles is the least lateral dimension with a ratio of 2:1 (L: B).

Table 5.4.3-5 Bearing capacities at Cable crossing

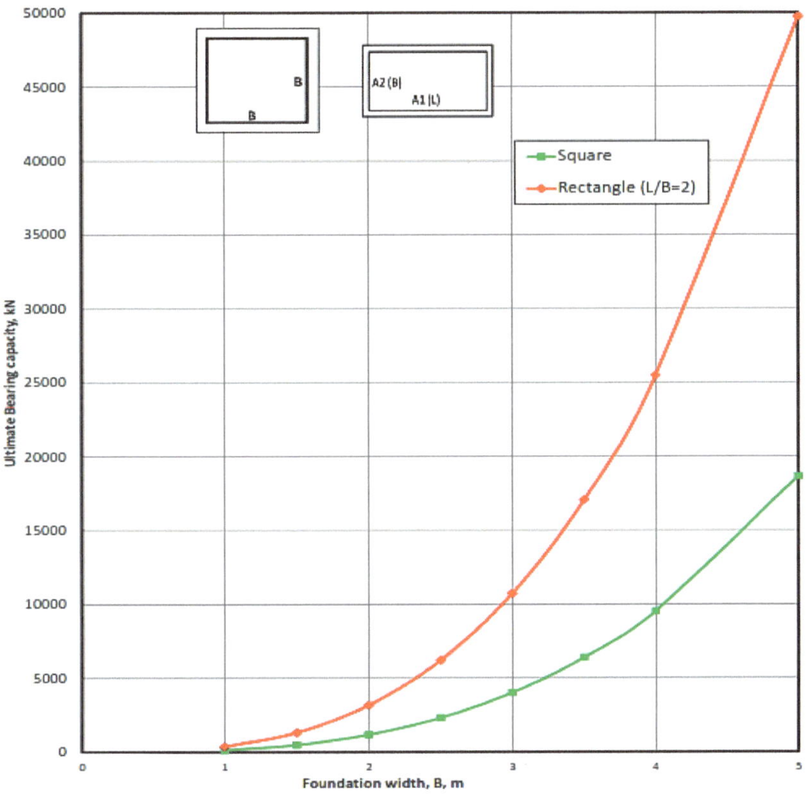

Figure 5.4.3-1: Typical Ultimate bearing capacity Vs Foundation width plot

Conclusions:

For the pipeline-soil interaction studies, geophysical and geotechnical information is extremely important in establishing proper zoning of soils along the pipeline route and thus arriving at appropriate ground model based on field and laboratory test results.

In addition, any recently conducted survey on the existing pipeline route in the nearby location in establishing the actual embedment and performance of pipeline is also very valuable information to arrive at the ground models. This case study illustrates a typical estimate of the ground model and the corresponding PSI parameters to be used for pipeline stability. Typical example of bearing capacity curves for shallow foundation is also presented.

This pipe route was successfully installed by the EPC contractor

5.5 Associated Risks and Engineering Solutions

The stratigraphy offshore Abu Dhabi majorly consists of cohesionless deposits which are carbonate in nature. In the seabed, there is a presence of either cemented or uncemented layers. In terms of pipeline design, if there is a presence of cemented layers at mudline, the pipes will not undergo any settlements upon laying and therefore would require higher thickness of coating to meet the stability requirements. In case of uncemented layers at mudline, the pile undergoes settlements due to its self-weight and with such deformations, it mobilizes some significant friction component which shall provide natural stability. The initial depth of 0.5m to 1m from the mudline shall have significant bearing on the design of the pipelines. Therefore, as a geotechnical engineer, one must accurately evaluate the nature and characteristics of the seabed. The proper zoning of the upper layers of the mudline is also of extreme importance. The significance of these findings shall have greater significance in terms of cost to provide the coatings on the pipelines. The same shall be true for drilled and grouted piles as well. In the region of Abu Dhabi, due to the presence of carbonate material, the presence of cemented layers of high thickness and significant strength at shallow depths are quite common. Therefore, the driven pile is NOT conducive to such stratigraphy, therefore, the commonly adopted method of piling operations is through the methodology of drilled and grouted piles.

The requirement of a sleeve pile is also decided based on the presence of the uncemented or cemented layers. The sleeve pile is required to avoid any possible borehole collapse. Therefore, it can be avoided in cases where the stratigraphy consists of cemented layers along the shallow depths of the borehole.

In cases of significant presence of uncemented layers along the shallow depths of the borehole, the option of sleeve pile is opted for during the design and implementation of a drilled and grouted pile.

Therefore, ground characterization is highly important for such engineering designs and executions. The footprints of the piles of a jacket structure shall be around 25m to 30m apart, the lateral continuity or discontinuity of the stratigraphy shall be extremely important to understand for the requirement of sleeve piles for the piling locations at each leg.

The lateral continuity of carbonate deposits is extremely variable and therefore these layers must be cautiously traced through a combination of geotechnical and geophysical site surveys.

In terms of Leg Penetration analysis of Jack ups rigs, due to the presence of cemented layers, the expected penetrations shall be shallow and would be generally less than 5m to 6m in this region. Therefore, scouring shall have a significant effect in such cases. During positioning, as the spud cans are extending towards a cemented mudline, there are chances of the spud cans getting damaged due to sudden movement. Therefore, rig moves are to be performed with extreme caution.

The design of mud mats usually is not very challenging considering the region of offshore Abu Dhabi. The likely challenges that could occur would be due to the uneven mudline stratigraphy which are to be closely studied and understood through geophysical techniques prior to designing the mud mats.

5.6 Lessons from the Experienced

5.6.1 From Survey Engineer to Industry Leader: Transforming Offshore Survey Services with Innovation and Vision – Mr. Tarek Abdulla Al Marzooqi.

Synopsis

The interview provides valuable insights into the evolution of offshore survey services and the critical role of precision in geotechnical investigations.

The journey from a small survey section to a standalone company reflects the industry's growing recognition of specialized survey services. Looking ahead, the focus on innovation, artificial intelligence, and unmanned solutions suggests a transformative future for offshore surveys, while maintaining the fundamental importance of accuracy and reliability in project execution.

Introduction

In an insightful conversation with a pioneering figure in offshore survey services, we explore the journey of transformation from a small survey section to a standalone company. This interview delves into the critical role of geotechnical investigations in offshore oil and gas projects, the importance of precision in survey services, and the vision for future innovations in the industry.

Interviewee Profile

Mr. Tarek Abdulla Al Marzooqi serves as the leader of Safeen Survey and Subsea Services, a company that emerged as a JV between AD Port Group and the National Petroleum Construction Company (NPCC); currently NMDC Group in 2022. With over two decades of experience in the offshore survey industry, he has been instrumental in transforming a small survey section into a comprehensive service provider. Starting his career as a survey-based engineer in 2001, he has led the development of integrated survey capabilities, combining geophysical, geotechnical, and ROV services.

Beyond his professional achievements, he maintains strong family values and continues his passion for football from his earlier days as a professional player.

The Talk

CRP: Dr CR Parthasarathy

TS: Mr. Tarek Abdulla Al Marzooqi, CEO Safeen Surveys and Subsea Services L.L.C.

CRP: Could you share your career journey and how you reached this leadership position?

TS: I started as a survey-based engineer at NPCC in 2001. The section was initially small, focusing only on geophysical surveying. We gradually built capabilities in geotechnical and other services to provide comprehensive solutions. After structuring the section as a profit center and operating for almost 20 years, our Shareholders established Safeen Survey and Subsea Services as a standalone company in 2022.

CRP: What significant experiences have shaped your career in this journey?

TS: Working under an EPC contractor presents unique challenges. We're involved throughout the project lifecycle, from pre-engineering surveys to as-built surveys. Precision is crucial as any mistakes can impact engineering design, installation, and procurement. The scope of geotechnical surveys has evolved significantly, expanding from just jacket installations to comprehensive requirements for cables, pipelines, and crossings.

CRP: How important is site characterization for risk assessment in offshore oil and gas construction?

TS: It's absolutely critical. While geotechnical work may be a small part of the budget in EPC contracts, inaccurate results can have severe consequences. We've experienced projects where jacket installation challenges arose due to imprecise geotechnical reports. The industry now recognizes this importance, leading to increased market size for geotechnical services.

CRP: What are Safeen's core principles and best practices?

TS: Our philosophy centers on delivering high-quality service and becoming an innovation hub. With 20 years of experience within NPCC, we understand market requirements and aim to provide different solutions than conventional approaches. We want to be a catalyst for new ideas and innovative methodologies.

CRP: What lessons have you learned from early career challenges?

TS: Learning through mistakes is essential. For instance, we initially thought we could use normal geotechnical drilling based on a DP2, but implementation showed we needed different solutions. Being part of an EPC contractor provided valuable learning opportunities, and every mistake contributed to our professional growth.

CRP: What is your vision for the future?

TS: We aim to become a one-stop company for geophysical, geotechnical, and subsea services. Our goal is to provide unique services and become an innovation hub, capable of operating at any depth, even 10,000 meters.

CRP: Could you share some insights about your personal life and daily routine?

TS: I'm very family-oriented and was previously a professional footballer, which I still enjoy weekly. My day starts with attending morning prayer, at the mosque, followed by work. I try to maintain physical activity after 5-6 PM, including walking and light sports. I have daughters and sons – with whom we maintain strong connections with our extended families, through regular weekend gatherings.

5.6.2 Leading Through Innovation: A Senior Executive's Journey in Gulf Energy Operations – Mr. Hesham Awda.

Synopsis

The interview provides valuable insights into successful leadership in the Gulf region's energy sector. Mr. Hesham Awda's emphasis on systematic work processes, continuous learning, and balanced approach to personal and professional life offers important lessons for industry professionals. His forward-looking perspective on market expansion and operational efficiency improvements reflects the dynamic nature of the energy sector and the importance of adaptability in corporate leadership.

Introduction

In this insightful interview, a senior executive shares his extensive experience in engineering and operations management across major Gulf region energy corporations. The conversation reveals valuable perspectives on managing mega-projects, developing business strategies, and the importance of systematic approaches to work in the modern energy sector.

His journey from engineering leadership to C-suite positions offers unique insights into successful corporate leadership in the Middle East's energy industry.

Interviewee Profile

The interviewee, Mr. Hesham Awda, is a seasoned energy sector executive with a distinguished career spanning engineering and senior management roles. He has served as VP of Engineering at Nokov Offshore, progressed to become VP for the Gulf region at NPCC, and currently holds the position of Chief Operating Officer at NMDC Energy.

His expertise includes managing mega-projects, particularly in Adnoc's field development initiatives, including NASAR, SARB, and other large scale projects, where he has been involved from conceptual stages through to operational implementation.

The Talk

CRP: Dr CR Parthasarathy

HA: Mr. Hesham Awda, Chief Operating Officer, NMDC Energy, Abu Dhabi.

CRP: Could you share the significant events or projects that shaped your career?

HA: Working on MEGA projects was a defining experience, particularly starting new fields in Adnoc. The involvement in NASAR, SARB, and several other large scale projects from conceptual stages through engineering, execution, and final operation was particularly significant. Such comprehensive project involvement is a rare opportunity in one's career. Additionally, the NPCC experience of developing business operations after signing the LTA (Long Term Agreement) was another crucial milestone.

CRP: What are the best practices and principles you adopt for success with large corporations in the region?

HA: Success relies on several key factors: ensuring the right assets that drive capabilities, capitalizing on resources and human capital, and investing in mission-critical assets and technologies. The business model focuses on implementing the right business strategy and entry strategy for different areas, followed by proper execution.

CRP: What challenges have you faced, and what lessons have you learned?

HA: Learning comes from experience, including operational challenges. The most important aspect is accident prevention. While small incidents may occur, what matters is our response, analysis, and systematic approach to prevention. Success and failure are part of life, but failures become valuable learning opportunities when taken seriously.

CRP: What tips would you share with our readers?

HA: The key takeaway from my experience is this: systemize your workflows. Too often, we get stuck firefighting—reacting to chaos instead of steering it. By systemizing, you create structure that handles the routine, freeing you to focus on what truly moves the needle: innovation, differentiation, and pushing boundaries. It’s about channeling your time and manpower into high-impact areas—efficiency upgrades, new ideas, digital transformation—while the system manages the rest. That’s how you stay ahead.

CRP: How have mentors shaped your leadership philosophy and values?

HA: From a national perspective, we deeply believe in our leaders. Sheikh Zayed, for instance, was a monumental role model whose guidance shaped how we approach leadership—how to manage crises, unite teams, and navigate complex situations. The same applies to Sherifa and Sheikh Muhammad, their ability to lead with clarity and vision has always inspired us. These figures set a standard for balancing authority with empathy, which we strive to embody. On a personal level, my father's influence has been foundational. He instilled the values of resilience and integrity, while my mother's decades-long career in education taught us the power of patience, mentorship, and fostering growth. Together, they've shaped how we operate as a family and as professionals—emphasizing unity and collective purpose.

CRP: What are your plans for the next 12 months and five years?

HA: The next 12 months will be very busy with numerous operator projects and expanding workload. We're focused on meeting expansion demands and serving client targets. Looking at the five-year horizon, plans include expanding operations, entering new markets, improving assets & capacity, and enhancing operational efficiency. We're particularly focused on gas projects, LNG projects, and strengthening partnerships.

CRP: Finally, could you share a glimpse into your daily routines, hobbies, and family life to round out our conversation?

HA: Regardless of whether I'm on leave or duty, I wake up early and embrace the day. Work is a source of enjoyment for me because I view it as a challenge. I prioritize spending quality time with my family daily, despite our many responsibilities and commitments. It's important to me to dedicate part of each day to connect with them and stay engaged in their lives. Health is equally vital: I ensure regular checkups and exercise as much as possible. My family is large—nine siblings in total, part of a close-knit extended family. I also have six children, now grown: three are engineers working in their fields, while the others are pursuing medicine or still in high school.

6
KINGDOM OF SAUDI ARABIA

6.1 Geological and Geotechnical aspects of Offshore Saudi Arabia

6.1.1. Introduction

Offshore Saudi Arabia, particularly the northern Persian Gulf region, is a globally significant area due to its vast oil and gas reserves. This chapter provides an in-depth analysis of the geological framework, stratigraphy, geotechnical characteristics, and engineering implications specific to this region. The focus is on presenting precise and actionable information for offshore geotechnical engineers working in this resource-rich but geologically complex environment.

6.1.2 Geological Framework

Tectonic Setting (Figure 6.1.2-1)

The offshore region of Saudi Arabia lies within the Arabian Basin, a tectonically stable intracratonic basin influenced by regional compressional forces from the Zagros orogeny. The Zagros Fold-Thrust Belt, located to the northeast, has indirectly shaped the structural features of the Arabian Basin through far-field stresses. Key structural trends include NW-SE anticlines formed during the Campanian period due to uplift and erosion events. These anticlines serve as major hydrocarbon traps, making them critical for exploration and production activities (Al-Dabbagh 2014).

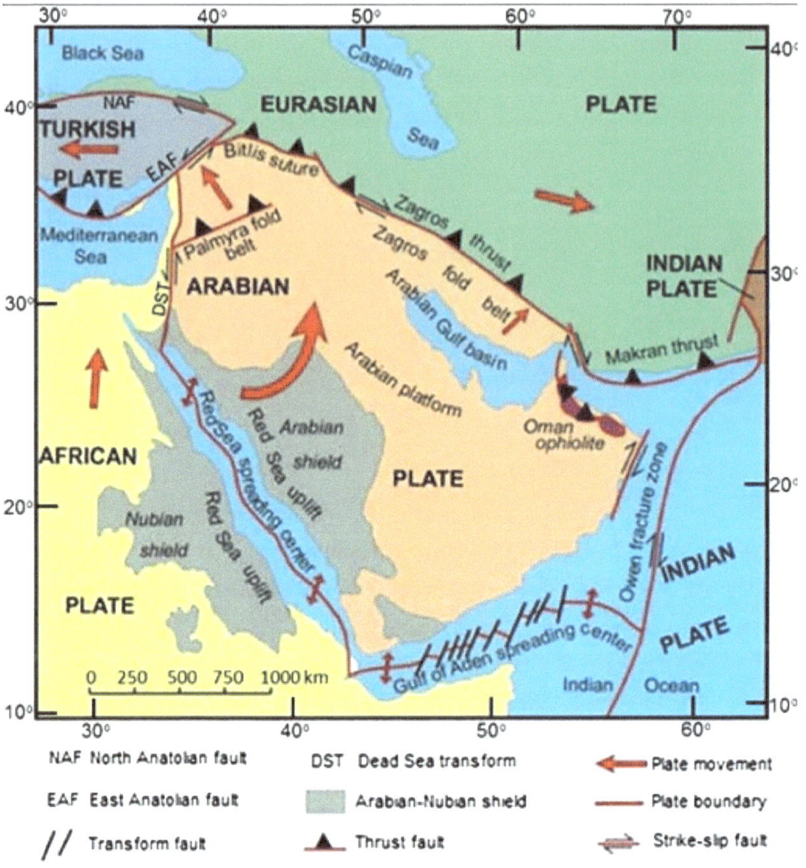

Figure 6.1.2-1: Arabian Plate with its boundaries (Al-Dabbagh 2014)

The tectonic stability of the Arabian Basin contrasts with the active deformation zones of the Zagros Mountains. However, localized fault systems, such as strike-slip faults and reverse faults, are present and influence reservoir compartmentalization and fluid migration pathways.

Stratigraphy

The stratigraphy offshore Saudi Arabia is dominated by Mesozoic formations, which host the majority of the region's hydrocarbon resources. Below is a detailed breakdown of key formations:

Jurassic Formations

1. Hanifa Formation: Composed of laminated lime mudstones rich in organic material, the Hanifa Formation serves as a primary

source rock for hydrocarbons. It contains type II kerogen, which generates high-quality oil upon maturation.

2. Tuwaiq Mountain Formation: This formation consists of limestone and dolomite sequences that act as secondary source rocks and seals. Its dense carbonate composition makes it an effective barrier for hydrocarbon migration.

Cretaceous Formations

1. Safaniya Member (Wasia Formation): The Safaniya Member is the main reservoir offshore Saudi Arabia. It comprises sandstone, siltstone, shale, and thin limestone layers. The sandstones exhibit excellent porosity (up to 30%) and permeability (ranging from 500–1,000 mD – millidarcy), making them ideal for hydrocarbon storage and flow.
2. Mauddud Formation : This formation acts as an intra-formational seal due to its dense carbonate composition. It effectively traps hydrocarbons within the underlying Safaniya reservoirs.

Petroleum Systems

The Mesozoic Petroleum System Offshore Saudi Arabia can be divided into two primary subsystems:

Jurassic Petroleum System

- Source Rocks: Organic-rich Jurassic formations like the Hanifa and Tuwaiq Mountain generate hydrocarbons.
- Seals: Evaporites, particularly anhydrite, provide impermeable barriers to trap hydrocarbons.
- Reservoirs: Carbonate platforms and sandstone units within the Jurassic sequence act as reservoirs.

Cretaceous Petroleum System

Source Rocks: Cretaceous shales and marls generate hydrocarbons, particularly in deeper basinal areas where thermal maturity is achieved.

- Reservoirs: The Safaniya Member within the Wasia Formation is the primary reservoir. Hydrocarbons migrate vertically and laterally into these sandstone units.

- Seals: Dense carbonates and evaporites within the Mauddud Formation act as seals.

Hydrocarbon generation and migration are driven by thermal maturation, which occurs at depths exceeding 4,000 feet. The water-driven mechanism in fields like Safaniya supports sustained production.

6.2 Geotechnical Characteristics

6.2.1 Subsurface Conditions

The subsurface geology offshore Saudi Arabia exhibits significant variability in rock types and mechanical properties. Key observations include:

1. Reservoir Depths: Hydrocarbon reservoirs are typically found at depths ranging from 4,000 to 7,000 feet within the Wasia Formation.
2. Rock Types:
   - Reservoir Rocks: Sandstones with high porosity (25–30%) and permeability (500–1,000 mD) dominate the productive zones.
   - Seal Rocks: Dense carbonates and evaporites, such as anhydrite, form impermeable barriers above reservoirs.

6.2.2 Soil Mechanics

The seabed sediments in the Persian Gulf exhibit variations in grain size and composition, influenced by depositional environments such as river deltas, marine currents, and wave action.

Common sediment types include:

1. Fine Silts and Clays: Predominant in deeper waters (>50 m) and characterized by low shear strength (<5 kPa) and high compressibility.
2. Sands and Gravels: Found near shorelines and in deltaic regions, these materials have higher shear strength and lower compressibility.
3. Calcareous Sediments: Composed of biogenic debris, these sediments are prevalent in carbonate-rich areas. Their mechanical behavior varies due to cementation effects.

Key geotechnical parameters for offshore foundations include:

- Shear Strength: Ranges from 2 - 5 kPa for soft clays to >50 kPa for stiff clays. Predominantly the cohesive deposits are heavily over consolidated in this region. Hard to Very Hard CLAYs are prevalent.
- Liquefaction Potential: Loose silty sands are susceptible to liquefaction during seismic events or wave-induced cyclic loading.
- Erodibility: Calcareous sediments and poorly consolidated clays are prone to erosion, posing risks to subsea infrastructure.

6.2.3 Engineering Implications

Foundation Design

Offshore structures, such as platforms and pipelines, require robust foundations to account for variable seabed conditions. Engineers must consider:

1. Sediment Stability: Fine-grained sediments may lose strength under cyclic loading from waves and currents.
2. Liquefaction Risks: Loose silty sands are prone to liquefaction during seismic events, necessitating stabilization measures.
3. Settlement Analysis: High compressibility of clayey sediments requires careful settlement predictions to avoid structural failure.

6.2.4 Drilling Challenges

Drilling operations face several challenges, including:

1. High-Pressure Zones: Deep reservoirs often contain overpressured zones, requiring advanced well control techniques.
2. Corrosion Risks: Saline environments accelerate corrosion rates, necessitating durable materials and protective coatings.
3. Shallow Gas: Trapped pockets of gas beneath the seafloor pose risks during drilling and require thorough site investigation.

6.3 Key Offshore Oil Fields

6.3.1 Safaniya Oil Field

The Safaniya Oil Field is the largest offshore oil field globally, spanning approximately 50 km by 15 km. It produces heavy crude oil with an API gravity of 27. Key characteristics include:

1. Reservoir Location: Reservoirs are located within the Wasia Formation at depths between 4,500–5,500 feet.
2. Production Mechanism: A strong water-driven mechanism supports production, ensuring sustained recovery rates.
3. Geological Features: The field's anticlinal structure and sealing lithologies contribute to its prolific hydrocarbon accumulation.

6.3.2 Marjan is an important offshore field in the Arabian Gulf. It is significant for its reserves of both oil and natural gas

6.3.3 Zuluf: Zuluf is another substantial offshore field in the Arabian Gulf. It consists of several smaller fields

6.3.4 Conclusion

Offshore Saudi Arabia represents a geologically intricate yet resource-abundant region.

The interplay of tectonic stability, Mesozoic stratigraphy, and hydrocarbon potential creates unique opportunities and challenges for offshore geotechnical engineering. By integrating detailed geological insights with advanced geotechnical methods, engineers can design safe, efficient, and resilient offshore infrastructure.

6.4 Introduction

Oil was first struck in the Arabian Peninsula at the Dammam No.7 well otherwise known as the "prosperity well" on March 3, 1938 (Oil & Gas News 2023). Thereafter, The Kingdom of Saudi Arabia has been the centerpiece of the oil market. Saudi Arabia's hydrocarbon correspondent, ARAMCO marked its 90 years of exploration in 2023. Saudi Arabia's largest offshore oil fields are Safaniya, Zuluf, Manifa, Marjan, and Abu Safah, which together hold a total 76.1 Bbbl in reserves with a combined production capacity of 3.42 MMb/d. These offshore fields represent roughly 30% of total Saudi reserves and 3.8% of daily global demand (Jongkees 2013).

6.5 General Geology of the Region

Anticlinal or domal structures in the sedimentary sequence of the north-eastern Arabian Platform and its offshore extension, contain all the known oil and gas fields of Saudi Arabia (ministry of Petroleum 1984). These currently comprise some fifty-six oil fields and four gas fields, all of which owe their origin to deep-seated tectonic movements in the Precambrian crystalline basement (Derakhshani and Farhoudi 2005).

In geology, basement and crystalline basement are crystalline rocks lying above the mantle and beneath all other rocks and sediments. They are sometimes exposed to the surface, but often they are buried under miles of rock and sediment.

Though there are significant talks and investments into renewable energy sources, hydrogen and carbon capture, oil remains the center of focus for the energy industry even at present. The renaissance offshore boom is being led by nationally owned oil companies in the Middle East. Saudi Arabian Oil Co. reported its plans to spend $55 billion in the coming year on capital investments, up from its previous plan of $35 billion for the next five years as announced in 2021. This is a significant increment in investment as it seeks to ramp up production from a series of fields in the Persian Gulf.

The top 5 offshore oil fields in Saudi Arabia are,

- **Safaniya Oil Field**: Safaniya is the largest offshore oil field in the world and is located in the Arabian Gulf. It was discovered in 1951 and has been in production since the late 1950s. Safaniya is operated by Saudi Aramco, the state-owned oil company of Saudi Arabia.
- **Manifa Oil Field**: Manifa is another major offshore oil field in Saudi Arabia, situated in the Arabian Gulf. It is notable for its heavy crude oil production. Saudi Aramco developed this field to increase its oil production capacity and offset declines in other fields.
- **Zuluf Oil Field**: Zuluf is another substantial offshore field in the Arabian Gulf. It consists of several smaller fields and is a significant contributor to Saudi Arabia's oil production.
- **Berri Oil Field**: Located in the Arabian Gulf, the Berri field is known for its high-quality Arabian light crude oil. It has been a key asset in Saudi Arabia's oil production and exports.
- **Marjan Oil Field**: Marjan is another important offshore field in the Arabian Gulf. It is significant for its reserves of both oil and natural gas. Saudi Aramco has initiated expansion projects to increase production capacity from Marjan.

The revenues from these fields have played a pivotal role in funding Saudi Arabia's economic and infrastructure development. The kingdom's oil policy and production decisions also have a significant impact on global energy markets, making Saudi Arabia a key player in the worldwide oil industry.

There are many features that are interesting about offshore oil & gas engineering such as the cutting-edge technology, the safety measures, and the geopolitical aspects involved.

The oil industry continually pushes the boundaries of technology. Offshore drilling rigs are marvels of engineering, with deep-sea installations that can extend thousands of feet below the ocean's surface. They use advanced robotics and artificial intelligence to optimize operations.

Some of the major Engineer, Procurement and Construction (EPC) contractors are,

1. Larsen & Toubro Hydrocarbon Energy
2. Subsea 7
3. NMDC Energy (Formerly NPCC)
4. McDermott
5. Saipem7
6. Lamprell

6.6 General Stratigraphies of Offshore Saudi Arabia

The following section provides a basic understanding of the typical stratigraphy of some the most important offshore oil and gas fields in Saudi Arabia. Refer **Table *6.6.1-1,* Table 6.6.2-1, Table 6.6.3-1 and Table 6.6.4-1 given below.**

6.6.1 Manifa Field

Depth (m)	Description	Shear Strength (Su), kPa,	UCS to Rock (MPa)	Angle of Internal Friction (phi)
0 – 3	Very Soft Sandy Carbonate CLAY	2 – 10	-	-
3 – 7	Loose to Medium Dense Calcareous Silica SAND	-	-	28
7 – 50	Dense to Very Dense Silica SAND	-	-	32
50 - 55	Hard Carbonate CLAY	200 - 300	-	-
55 – 58	Weak to Medium Cemented CALCISILTITE	-	10 – 15	-
58 – 60	Weak to Medium Cemented CALCARENITE	-	10 – 15	-
60 – 68	Very Weak to Weak CALCARENITE	-	5 – 10	-
68 – 75	Very Weak to Weak CALCISILTITE	-	5 – 10	-

Table 6.6.1-1: Generalized Soil Stratigraphy of Manifa Field

6.6.2 Marjan Field

Depth (m)	Description	Shear Strength (Su), kPa,	UCS of Rock	Angle of Internal Friction (phi)
0 – 1.5	Very Soft Sandy Calcareous CLAY	2 – 5	-	-
1.5 – 3	Very Dense Silty Calcareous Silica SAND	-	-	34
3 – 4.5	Very Stiff to Hard Sandy Calcareous CLAY	150 – 250	-	-
4.5 – 10	Dense to Very Dense Calcareous Silica Very Silty SAND	-	-	33
10 – 20	Hard Calcareous CLAY	250 – 500	-	-
20 – 27	Very Dense Very Silty Calcareous Silica SAND	-	-	34
27 – 30	Hard Calcareous SILT	250 – 300	-	-
30 – 37	Hard Calcareous CLAY	300 – 500	-	-
37 – 40	Very Dense Silty Calcareous Silica SAND	-	-	34
40 – 45	Hard Calcareous CLAY	300 – 500	-	-
45 – 55	Hard CLAY	250 – 500	-	-
55 – 75	Very Hard Calcareous CLAY	400 – 700	-	-

Table 6.6.2-1: Generalized Soil Stratigraphy of Marjan Field

6.6.3 Zuluf field

Depth (m)	Description	Shear Strength (Su), kPa,	UCS of Rock	Angle of Internal Friction (phi)
0 – 2	Very Soft Sandy Calcareous CLAY	2 – 10	-	-
2 – 3	Firm to Stiff Calcareous CLAY	25 – 75	-	-
3 – 5	Stiff to Very stiff Calcareous CLAY	75 – 120	-	-
5 – 5.5	Very Dense Calcareous Silica SAND	-	-	34
5.5 – 6.5	Very Stiff Sandy Calcareous CLAY	125 - 175	-	-
6.5 – 7.5	Stiff to Very Stiff Carbonate CLAY	100 – 150	-	-
7.5 – 10	Dense to Very Dense Silica SAND	-	-	33
10 – 10.5	Hard Sandy Calcareous CLAY	250 – 500	-	-
10.5 – 20	Very Dense Calcareous Silica SAND	-	-	34
20 – 22	Extremely Weak to Very Weak GYPSUM	-	3 – 5	-
22 – 28	Very Stiff to Hard Sandy Calcareous CLAY	200 – 300	-	-
28 – 35	Very Dense Calcareous Silica SAND	-	-	34
35 - 40	Hard to Very Hard CLAY	250 – 500		-
40 – 70	Hard Calcareous CLAY	200 – 300	-	-
70 - 75	Very weak to weak GYPSUM	-	5 - 20	-

Table 6.6.3-1: Generalized Soil Stratigraphy of Zuluf Field

6.6.4 Safaniya Field

Depth (m)	Description	Shear Strength (Su), kPa,	UCS of Rock	Angle of Internal Friction (phi)
0 – 0.5	Very Loose Silica SAND	-	-	25
0.5 – 1.5	Very Soft Sandy Carbonate CLAY	2 – 5	-	-
1.5 – 2.5	Very Loose to Loose Calcareous Silica SAND	-	-	26
2.5 – 5	Medium Dense Calcareous Silica SAND	-	-	28
5 – 7	Dense to Very Dense Calcareous Silica SAND	-	-	32
7 – 8.5	Stiff to Hard CLAY	50 – 250	-	-
8.5 – 32	Very Dense Silica SAND	-	-	34
32 – 35	Very Stiff to Hard Sandy CLAY	150 – 300	-	-
35 – 39	Very Dense Silty Silica SAND	-	-	34
39 – 45	Hard Sandy CLAY	200 – 500	-	-
45 – 56	Very Dense Silty Silica SAND	-	-	34
56 – 58	Hard Sandy CLAY	200 – 500	-	-
58 – 75	Dense to Very Dense Silty Silica SAND	-	-	32

Table 6.4.4-1: Generalized Soil Stratigraphy of Safaniya Field

6.7 Case Studies

6.7.1 Case Study-1: Vibration Monitoring (Dr Prashanth Talkad, Ms. Minnu, Ashitha, Srinivas Nagappa and Nandagopal)

Problem Statement

Pile driving causes vibrations in the nearby structure. The vibration due to pile driving must be within the allowable limit for safety. The pile-driving-induced vibrations can be monitored by instrumenting the nearby structure with accelerometers. Generally, the peak particle velocity induced by vibration is restricted to being less than 50mm/sec for structures, beyond which damage is expected. This case study deals with instrumenting the nearby jetty structure and monitoring the vibrations caused by pile driving during a new six-leg Jacket structure installation in an offshore field in Saudi Arabia. It was required to

a) Ensure the safety of the existing Jetty structure during the installation of the Jacket by understanding the potential reduction in foundation capacity and
b) Evaluate if any additional restraints must be provided to prevent slip of concrete girders due to excessive movement or vibration of the structure.

Project Details:

A six-legged Jacket structure with 42" (1.067m) x 1.5" driven piles & 50m design penetration was to be installed nearby to an existing Jetty

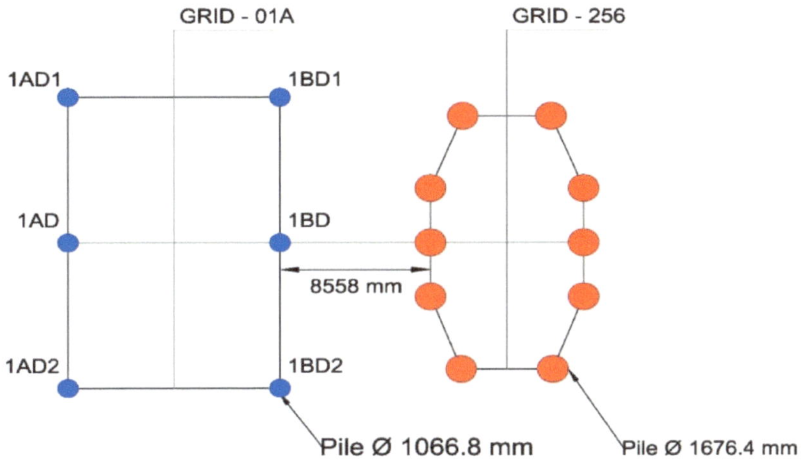

Figure 6.7.1-1: Six-Leg Jacket and the existing Jetty (trestle)

(trestle) structure supported by 10 number of piles of 66" (1.676m) outer diameter with 11 m penetration, supporting concrete girders carrying active pipelines.

The Row B piles of the jacket were close to the Jetty and the center-to-center distance between the Row B piles and the existing trestle piles was 8.558m (>8D of the Jacket's main piles) as shown above in **Figure 6.7.1-1.**

Though multiple measures were suggested from our end as the consultant, due to several factors, it was decided that vibration monitoring should be conducted throughout the driving procedures of the Jacket's main piles.

Vibration Monitoring:

Strain gauges and accelerometers were affixed diagonally across from each other on the girder nearest to where the main piles were being driven. Vibration monitoring on Jetty was conducted during the entire duration of driving of the main piles of the Jacket. The driving of the Jacket piles to the target penetration was accomplished with a combination of MHU-300S and MHU-800S hammers.

Based on vibration monitoring on jetty, no data was recorded during the driving of P1 segments with the MHU-300S hammer and thus the MHU-300S hammering of the P1 segments of all main piles did not significantly affect the trestle in terms of vibration generation. During the driving of 1BD2 and 1BD piles, refusal occurred at depths of 22.0m and 29.25m respectively while using the MHU-300 hammer, necessitating the switch to the MHU-800 hammer for further driving.

During the intense driving phase, vibrations were recorded on the trestle between depths of 22.0m and 23.5m for 1BD2 and at 29.25m for 1BD. Additionally, vibrations were detected during the driving of P2 segments of all piles using the MHU-800 hammer, with recorded velocity values generally within the 15mm/sec limit set by the client. Occasional exceedances of up to 18mm/s were observed, particularly during the closest pile's (1BD) driving. These excessive vibrations were mitigated by reducing hammer energy. It's important to note that not every hammer blow induced measurable vibrations on the trestle due to its distance of at least 8.558m from the hammering location, connected solely through the soil medium. All recorded vibrations on the structure have been continuously monitored and are presented in **Table 6.7.1-1.**

Overall, based on the monitoring data, it can be concluded that the driving of the main piles of the six-legged jacket platform did not adversely affect the existing trestle structure.

Pile	Monitoring Depth (m)		Hammer Blows	Blows Triggering Vibration on Trestle	Velocity Range (mm/s)	Remarks
	From	To				
ROW 1A - Farther Piles						
1AD-P2	31.13	50.05	2276	201	4 - 14	Hard driving due to high setup – driven after weather standby.
1AD1-P2	31.01	50.01	2420	238	2 - 15	
1AD2-P2	31.01	49.86	2568	36	5 - 12	-
1AD2 Restrike	49.86	49.86	252	41	9 - 12	-
ROW 1B - Closer Piles						
1BD-P1	29.22	29.31	676	97	3 – 18	Refusal Pile – Higher values observed during hard driving.
1BD1-P2	31.13	50.05	1963	222	6 - 16	2 – 5 blows of higher values observed during hard driving.
1BD2-P1	22.16	27.44	1978	156	4 - 15	Refusal Pile – Higher values observed during hard driving.

Table 6.7.1-1: Vibration monitoring details-MHU 800S hammer

Conclusions

Pile-driving-induced vibrations can be monitored by instrumenting the nearby structure with accelerometers. The pile driving during the installation of Jacket piles induced no vibration when the driving was done using the MHU 300S hammer due to low energy imparted during driving. However, the pile driving using the MHU 800S hammer generated measurable vibrations, with a maximum velocity that was below the recommended limit of 15mm/s. Where the peak particle velocity exceeded, the hammer energy was adjusted to restrict the same to be less than the prescribed limit.

This project was successful in meeting the objectives. Overall, based on the monitoring data, it can be concluded that the driving of the main piles of the six-legged jacket platform did not adversely affect the existing trestle structure.

6.7.2 Case Study-2 (Dr CR Parthasarathy, Dr Ramesh Gangisetti, Ms. Minnu, Ms. Swathi, Mr Srinivas Nagappa & Mr Kris Gustavian)

Problem Statement

Of late, installation of monopiles is carried out by using hydraulically operated or electrically operated Vibro hammer with the biggest advantage being able to position/pin the pile and control the alignment during installation. The other advantages of hydraulic vibro hammers are to achieve better control during piles run down and induces less stress on the pile. The hydraulic vibrohammer market has expanded in leaps and bounds over the last few years to meet the demand.

Using vibrohammer, the monopiles are initially firmly pinned or installed till they reach the desired penetration. Then these monopiles are driven further to the desired penetration or strike the pile (when already designed penetration is achieved) using an impact hammer. Where the-built capacity assessment is required, the monopile is instrumented with strain gages and accelerometers and monitored during the impact driving to capture a force-velocity signature for estimating the available shaft and toe capacity using CAPWAP.

This case study deals with monopiles of Large and Small Berthing Platform, installed initially using a vibratory hammer up to certain penetrations followed by an impact hammer driving until the final penetration was achieved.

Concern was raised about the lateral pile capacity based on the assumption that the installation using vibro hammer causes greater soil strength degradation than impact driving. It was demonstrated that pile monitoring data of impact driving can be used to estimate the remolded soil strength parameters for the depths where the pile was installed using vibratory hammer. The remolded soil strength and associated parameters can then be used for pile-soil interaction modeling in PLAXIS-3D for evaluating the lateral pile capacity. It was concluded that vibratory hammer installation does not cause complete degradation of soil strength, as was apprehended.

Project Details

At an offshore field of the Kingdom of Saudi Arabia, monopiles, 4 Nos of 4m outer diameter for Large Breasting Dolphin (LBD) and 10 Nos of 2m outer diameter for Small Breasting Dolphin (SBD), were planned to be installed to final penetrations of 32m to 40m for LBD monopiles and 28m for SBD monopiles respectively. These monopiles were initially installed using a vibratory hammer to penetrations 10.5 to 22.5m (LBD) and 15.5 to 20.25m (SBD) followed by an impact hammer driving up to final penetrations. One of the LBD Monopile BP03 refused at 29.2m.

The key plan of the LBD and SBD monopiles is shown in **Figure 6.7.2-1.** The monopiles geometry, hammers used, and penetrations achieved are summarized in and **Table 6.7.2-1 and Table 6.7.2-2** for LBD and SBD Piles respectively.

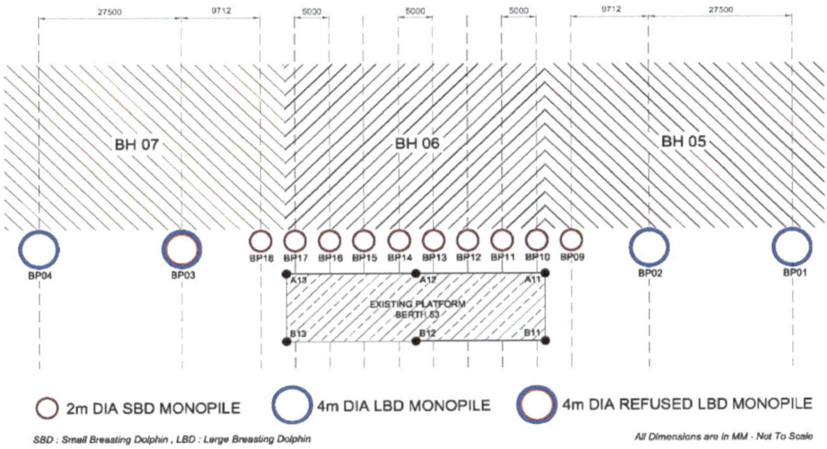

Figure 6.7.2-1: Key plan of SBD and LBD monopiles

**DMS – Dynamic Monitoring System*

LBD Piles OD 4.0m Driving Details							
Sl. No.	Pile ID	Total Pile Length |LE| (m)	Target Pen (m)	Pile + ILT (m)	Driven with Vibro Hammer (m)	Driven with Impact Hammer MHU 1900 with DMS* (m)	Remark
1	BP1	69.4	40.00	3.75	3.75 to 22.5	22.5-40.00	3.5 hrs of delay at 32.0 depth
2	BP2	69.5	40.00	4.00	4.00 to 18.5	18.5 - 40.00	6.0 hrs of delay at 32.0 m depth
3	BP3	60.5	31.00	5.00	5.00 to 10.50	10.50-29.10	Refused at 29.1m with S-1900 hammer
4	BP4	60.4	31.00	4.25	4.25 to 17.50	17.50-31.00	

Table 6.7.2-1: LBD monopiles details

**DMS – Dynamic Monitoring System*

SBD Piles OD 2.0m Driving Details						
Sl. No.	Pile ID	Total Pile length |LE| (m)	Target Penetration (m)	Pile + ILT (m)	Driven with Vibro Hammer (m)	Driven with Impact MHU 1900 Hammer with DMS* (m)
1	BP09	57.6	28	4.00	400 to 20.00	20.00 to 28.00
2	BP10	57.35	28	3.75	3.75 to 20.00	20.00 to28.00
3	BP11	57.45	28	4.00	4.00 to 20.00	20.00 to 28.00
4	BP12	57.35	28	4.25	4.25 to 20.00	20.00 to 28.00
5	BP13	57.35	28	4.50	4.50 to 20.25	20.00 to 28.00
6	BP14	57.25	28	4.75	4.75 to 20.00	20.00 to 28.00
7	BP15	57.25	28	4.75	4.75 to18.75	18.75 to 28.00
8	BP16	57.25	28	5.00	5.00 to18.50	18.75 to 28.00
9	BP17	57.35	28	4.75	5.00 to 15.50	15.50 to 28.00
10	BP18	58.0	28	4.75	4.75 to15.75	15.75 to 28.00

Table : 6.7.2-2: SBD monopiles details

Soil Conditions

As shown in **Figure 6.7.2-1** (above), the SBD monopiles installed were close to site-specific borehole BH-06. The seabed soils consist of loose to medium dense carbonate SAND layers up to 6.0m. The soil then is predominantly very stiff to Hard CLAY up to a depth of 47.8m with two in-between SILT layers of 2.55m thick at 7.3m and 6.1m thick at 14.3m and a very dense cemented SAND layer at 22m to 24.3m. The design soil parameters for axial pile capacity are shown in **Table 6.7.2-3.**

Layer	Soil/ Rock Type	Depth (m)		Submerged Unit Weight γ' (kN/m³)	Shear strength, Su kPa		$CaCO_3$ (%)	API RP 2GEO (2014)			
		From	To		From (top)	To (base)		δ Soil-Pile Friction Angle (θ)	Bearing Capacity Factor Nq (-)	Limiting Skin Friction f_{lim} (kPa)	Limiting Unit End Bearing q_{lim} (kPa)
1	SILT	0.00	0.30	7.20	3	3	-	-	-	-	-
2	SAND	0.30	0.65	8.30	-	-	90	10.00	8	20.0	1900
3	SAND	0.65	4.70	8.70	-	-	87	17.50	16	20.0	2725
4	SAND	4.70	6.00	10.30	-	-	16	27.50	30	125.0	10750
5	CLAY	6.00	7.3	10.30	250	250	-	-	-	-	-
6	SILT	7.30	9.85	7.95	165	165	-	-	-	-	-
7	CLAY	9.85	12.8	9.80	400	400	-	-	-	-	-
8	CLAY	12.80	14.3	9.10	225	225	-	-	-	-	-
9	SILT	14.30	20.4	9.00	225	225	-	-	-	-	-
10	CLAY	20.40	22.0	9.00	460	460	-	-	-	-	-
11	SAND	22.0	24.3	8.45	-	-	86	30.0	50	20.0	3000
12	CLAY	24.3	30.0	9.5	430	430	-	-	-	-	-
13	CLAY	30.0	33.1	9.75	280	280	-	-	-	-	-
14	CLAY	33.1	39.8	7.8	400	400	-	-	-	-	-
15	CLAY	39.8	44.9	7.8	205	205	-	-	-	-	-

Table 6.7.2 3: Soil design parameters for axial compression capacity (SBD)
(BH-06, Ref Figure 6.7.2-1)

The Unit Skin Friction (USF) and Unit End Bearing (UEB) vs. Pile penetration, as well as the Axial compression and tension capacity vs. Pile penetration, calculated as per **API RP2GEO-2011 (R2021)**, are presented in **Figure 6.7.2-2, Figure 6.7.2-3** and **Figure 6.7.2-4** respectively.

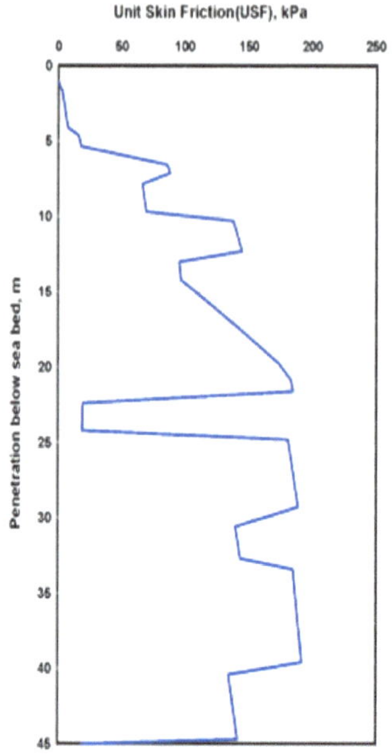

Figure 6.7.2-2: USF vs. Pile penetration (SBD)

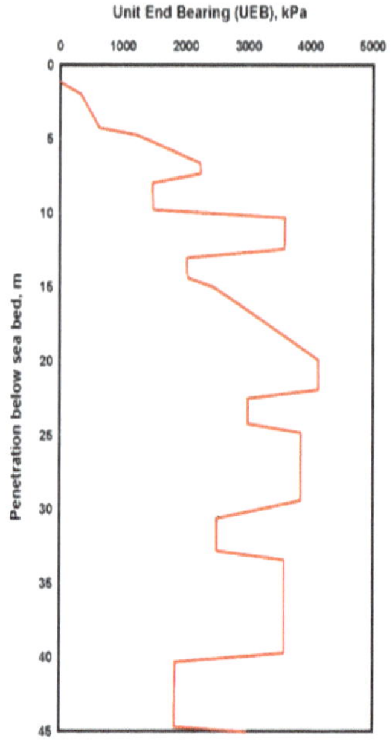

Figure 6.7.2-3: UEB vs. Pile penetration (SBD)

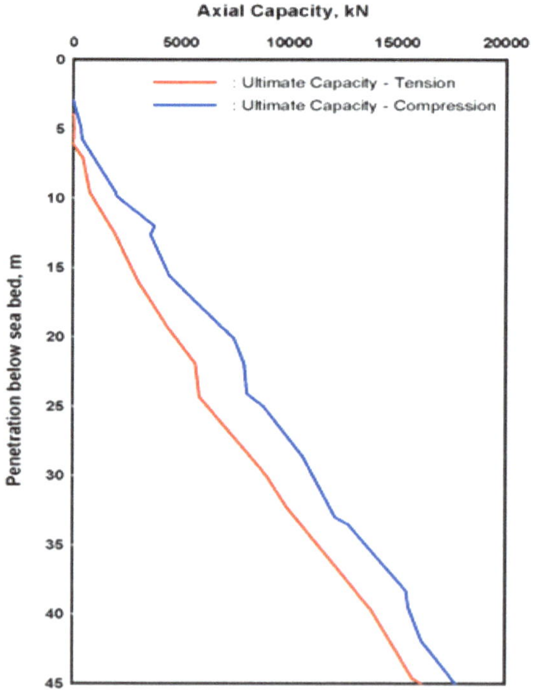

Figure 6.7.2-4
Axial Monopile Compression & Tension Capacity – 2m OD SBD (as per BH-06)

Similarly, the geotechnical data of BH-5 and BH-7 (**ref Figure 6.7.2-1 above**) is presented in **Table 6.7.2-4 and Table 6.7.2-5** respectively. Based on these parameters, the axial capacity of 4.0m Diameter monopile can be estimated as shown in **Figure 6.7.2-5 and Figure 6.7.2-6.**

Layer	Soil/ Rock Type	Depth (m)		Submerged Unit Weight Y' (kN/m3)	Shear strength, Su kPa		CaCO3 (%)	API RP 2GEO (2014)			
		From	To		From (top)	To (base)		δ Soil-Pile Friction Angle (deg)	Bearing Capacity Factor Nq (-)	Limiting Skin Friction flim (kPa)	Limiting Unit End Bearing qlim (kPa)
1	SILT	0	0.15	7.8	2	2	-	-	-	-	-
2	SAND	0.15	4.1	8.6	-	-	81	12.5	10	20	2175
3	SAND	4.1	9.4	10.2	-	-	81	27.5	45	20	3000
4	SILT	9.4	10.25	9.9	100	100	-	-	-	-	-
5	SAND	10.25	12.2	9.4	-	-	86	27.5	45	20	3000
6	CLAY	12.2	17.75	9.9	170	170	-	-	-	-	-
7	SILT	17.75	19	9.8	165	165	-	-	-	-	-
8	CLAY	19	25.1	9.9	90	175	-	-	-	-	-
9	SAND	25.1	25.5	9.6	-	-	86	25	40	20	3000
10	CLAY	25.5	33.4	10	180	275	-	-	-	-	-
11	SILT	33.4	35.75	9.5	185	185	-	-	-	-	-
12	SAND	35.75	39.6	10.4	-	-	86	27.5	45	20	3000
13	SILT	39.6	41.8	9.4	300	300	-	-	-	-	-
14	CLAY	41.8	45.2	8.5	300	300	-	-	-	-	-

Table 6.7.2-4: Soil design parameters for axial compression capacity (LBD) (BH-05 – Ref 6.7.2-1)

Layer	Soil/ Rock Type	Depth (m)		Submerged Unit Weight Y' (kN/m3)	Shear strength, Su kPa		CaCO3 (%)	API RP 2GEO (2014)			
		From	To		From (top)	To (base)		δ Soil-Pile Friction Angle (deg)	Bearing Capacity Factor Nq (-)	Limiting Skin Friction flim (kPa)	Limiting Unit End Bearing qlim (kPa)
1	SILT	0	0.15	7.9	2	2	-	-	-	-	-
2	SAND	0.15	1.1	9	-	-	90	10	8	20	1900
3	SAND	1.1	4.85	9.3	-	-	90	12.5	10	20	2175
4	SAND	4.85	5.7	10.15	-	-	83	30	50	20	3000
5	CLAY	5.7	6.6	10.15	300	300	-	-	-	-	-
6	CLAY	6.6	10.3	10.5	125	180	-	-	-	-	-
7	SAND	10.3	11.1	9.5	-	-	74	27.5	30	23.9	3253
8	CLAY	11.1	14.6	10.05	320	320	-	-	-	-	-
9	SILT	14.6	21.1	9	320	400	-	-	-	-	-
10	CLAY	21.1	24.1	7.8	400	400	-	-	-		-
11	SAND	24.1	25.3	9.3	-	-	77	32.5	45	-22.4	3221
12	CLAY	25.3	31.1	9.8	360	360	-	-	-	-	-
13	CLAY	31.1	37	9.1	500	500	-	-	-	-	-
14	SILT	37	40	9.8	400	400	-	-	-	-	-
15	SILT	40	44.1	7.9	360	360	-	-	-	-	-
16	Gypsum	44.1	45.1	11.2	-	-	-	25	20	81	12000

Table 6.7.2-5: Soil design parameters for axial compression capacity (LBD) (BH-07 Ref Figure 6.7.2-1)

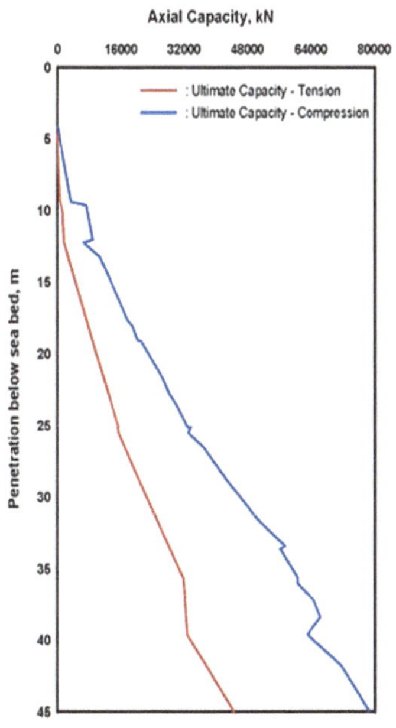

Figure 6.7.2-5 : Axial monopile compression and tension capacity 4m OD, LBD (as per BH-05)

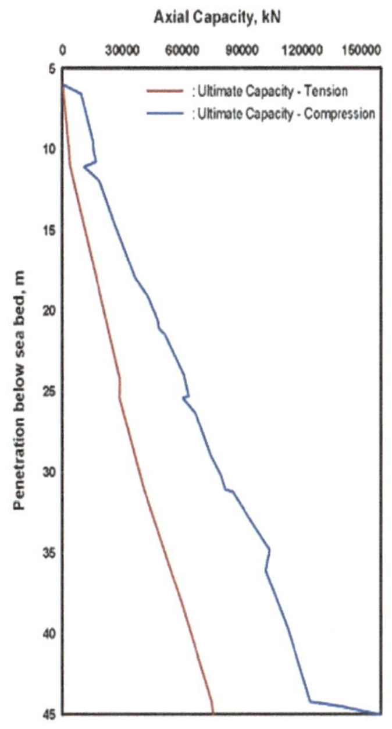

Figure 6.7.2-6 : Axial monopile compression and tension capacity 4m OD, LBD (as per BH-07)

Pile monitoring

After initial penetration was achieved by Vibro hammer, Impact driving was resorted to. All monopiles were instrumented and pile monitoring was carried out during impact driving. The pile monitoring details are summarized in **Table 6.7.2-6**. The CAPWAP estimated compression capacities at the end-of-driving (EOD) and the FOS achieved for operating and extreme storm conditions are summarized in **Table 6.7.2-7.**

BP-01 and BP-02 (4.0m dia) Monopiles were driven to 22.50m and 18.5m using vibrohammer respectively. Impact hammer was utilized to drive from 22.5m and 18.5m onwards till 40m penetration depth. However, the mono piles were driven to a depth of 32m without pile monitoring measurements due to difficulty in accessing the pile top to mount the sensors. Upon reaching 32m, the sensors were rigged up. During this activity, there was a delay in impact driving for 3.5hr and 6hr respectively for BP-01 and BP-02 Monopiles. This delay in driving can be considered to calculate mobilized setup factor.

The capacity mobilized at End of Driving (EOD-32m) was estimated based on Hind cast analysis, while the Beginning of Driving (BOD) capacity is computed from CAPWAP analysis.

Pile Number	**Pile monitoring depths, m**
	From-To
	SBD Piles
BP09	20.00 - 28.00
BP10	20.00 - 28.00
BP11	20.00 - 28.00
BP12	20.00 - 28.00
BP13	20.25 - 28.00
BP14	20.00 - 28.00
BP15	18.75 - 28.00
BP16	18.50 - 28.00
BP17	15.50 - 28.00
BP18	15.75 - 28.00
LBD Piles	
BP1	32.00 - 40.00
BP2	32.00 - 40.00
BP3	10.50 - 29.10
BP4	17.50 - 28.00

Table 6.7.2-6: Summary of pile monitoring depth

SBD	Driving Condition	Vertical penetration of Analysis	PILE CAPACITY (CAPWAP)			Calculated Factors of Safety							
						In-Place				Mooring			
						Operating		Extreme		Operating		Extreme	
			Shaft Resistance	End Bearing	Total Pile Capacity	Unfactored Load	Calculated FOS (FOS > 2)	Unfactored Load	Calculated FOS (FOS > 1.5)	Unfactored Load	Calculated FOS (FOS > 2)	Unfactored Load	Calculated FOS (FOS > 1.5)
		(m)	(MN)	(MN)	(MN)	(MN)		(MN)		(MN)		(MN)	
Monopiles of 2m OD													
BP9	EOD	28	9.26	2.83	**12.09**	3.17	**3.81**	3.43	**3.52**	2.52	**4.8**	2.92	**4.14**
BP10		28	9	3.16	**12.15**	2.42	**5.02**	2.68	**4.54**	2.21	**5.5**	2.47	**4.93**
BP11		28	7.38	3.96	**11.33**	2.13	**5.33**	2.38	**4.76**	2.33	**4.87**	2.57	**4.4**
BP12		28	8.89	3.46	**12.35**	2.13	**5.81**	2.37	**5.2**	2.59	**4.78**	2.84	**4.35**
BP13		28	9.46	2.88	**12.34**	2.08	**5.95**	2.33	**5.3**	2.32	**5.32**	2.57	**4.79**
BP14		28	9.24	4.41	**13.65**	2.07	**6.59**	2.32	**5.88**	2.32	**5.88**	2.57	**5.3**
BP15		28	9.3	3.81	**13.11**	2.15	**6.1**	2.4	**5.46**	2.6	**5.04**	2.85	**4.59**
BP16		28	12.26	4.29	**16.56**	2.07	**8**	2.33	**7.12**	2.32	**7.15**	2.57	**6.45**
BP17		28	16.7	14.81	**31.51**	2.23	**14.13**	2.49	**12.67**	2.42	**13.02**	2.67	**11.78**
BP18		28	20.2	26.07	**46.27**	2.85	**16.24**	3.11	**14.86**	3.17	**14.6**	3.43	**13.48**
Monopiles of 4m OD													
BP1	EOD	40	24.5	7.51	32.01	7.62	4.2	8.84	3.62	10.68	3	11.91	2.69
BP2		40	24.71	9.5	34.21	7.62	4.49	8.84	3.87	10.68	3.2	11.91	2.87
BP3		29.1	44.51	65.95	110.47	8.36	13.21	9.39	11.77	7.55	14.64	9.32	11.85
BP4		31	20.73	10.57	31.29	8.36	3.74	9.39	3.33	7.55	4.15	9.32	3.36

Table 6.2.7-7: CAPWAP estimated compression capacities-Monopiles of 2m OD

Interpretation of as-installed soil shear strength parameters

The soil around the pile surface will get remolded due to continuous driving and lose strength. Therefore, to determine the available lateral pile capacity at the time of driving, the as-installed shear strength parameters need to be evaluated.

At the end of driving, the USF from CAPWAP represents the as-installed soil conditions and thus the as-installed strength parameters. These measured parameters account for the effects of both Vibratory and Impact Hydraulic hammers. The USF values estimated from CAPWAP were used to back-calculate the shear strength parameters, undrained cohesion Cu, for clay layers, and effective angle of internal friction ϕ' for sand layers using the procedure given below.

For cohesive soils

The unit shaft friction, f_s, is as follows:

$$f_s = \alpha S_u \quad \textit{Equation 6.7.2-1}$$

where,

α= Adhesion factor (dimensionless)

S_u= Undrained shear strength (kPa)

From the author's experience and database, α ranges between 0.3 to 0.6 in similar soils. Hence for the current analysis, a value of 0.5 is assumed throughout the pile penetration depths.

For example:

As per the GT report, for the Clay layer encountered at a depth of 12.0m, the CAPWAP USF obtained is 49.85kPa, hence the shear strength back calculated is 49.85/0.5=99.7kPa.

For cohesionless soils

For non-calcareous siliceous cohesionless soils, the unit shaft friction, f_s is defined as:

$$f_s = \beta \sigma'_v \quad \textit{Equation 6.7.2-2}$$

where, β = shaft friction factor (dimensionless)=k tan δ

σ'_v = effective vertical stress at the point of interest (kPa)

For example:

As per Geotechnical investigation report, for a Sand layer which is encountered at a depth of 11.0m, the effective over burden pressure is 102.88kN/m^3, while the CAPWAP USF obtained is 22.05Kpa. Hence the delta value back calculated is $\tan^{-1}(22.05/(0.8*102.88)) = 15^{\circ}$ and Phi' = delta+5deg.

The back-calculated soil setup values from the EOD and BOD CAPWAP capacities and the setup measured from a re-strike test conducted on a nearby pile of the Jacket Berth-53 are summarized in **Table 6.7.2-8.**

Location	Setup Hrs	Setup Factor	Remark
BP1	3.5hr	1.35	Vibro + Impact
BP2	6hr	2.59	Vibro + Impact
B53	12hr	1.86	Impact

Table 6.7.2-8 Soil setup calculated from EOD and BOD capacities of monopiles and EOD and Re-strike capacities of the Jacket Berth-53

The CAPWAP estimated USF vs. Pile penetration for SBD and LBD monopiles for EOD (@Target Penetration) are shown in **Figure 6.7.2-7 and Figure 6.7.2-8.**

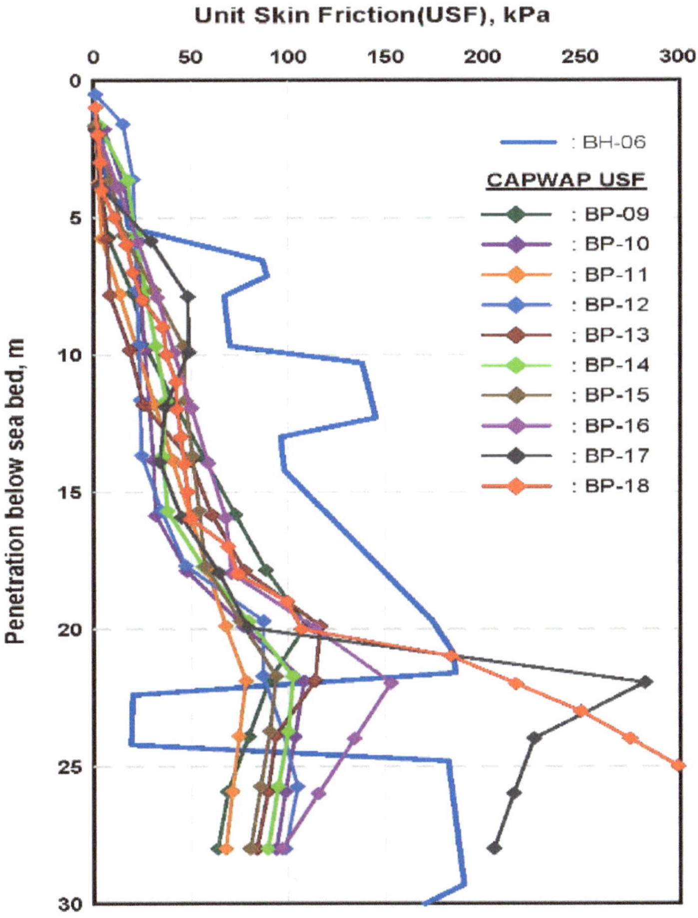

Figure 6.7.2.7: Unit Skin Friction (USF) vs. Pile penetration-EOD (SBD, 2m OD Monopiles)

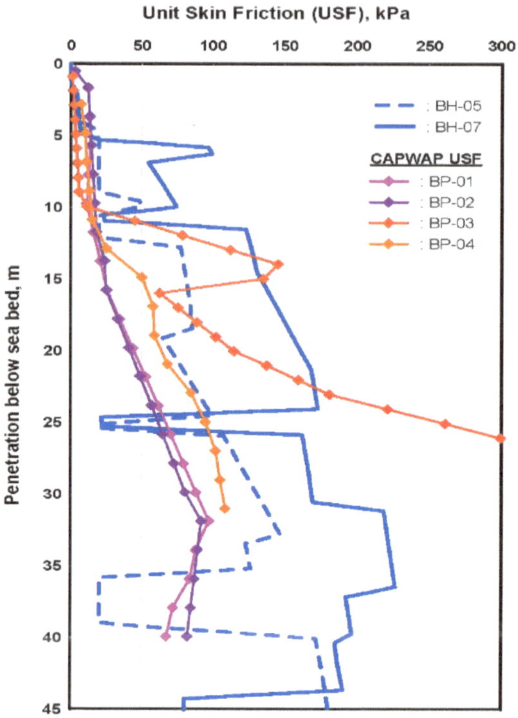

Figure 6.7.2-8: Unit Skin Friction (USF) vs. Pile penetration-EOD (LBD, 4m OD Monopiles)

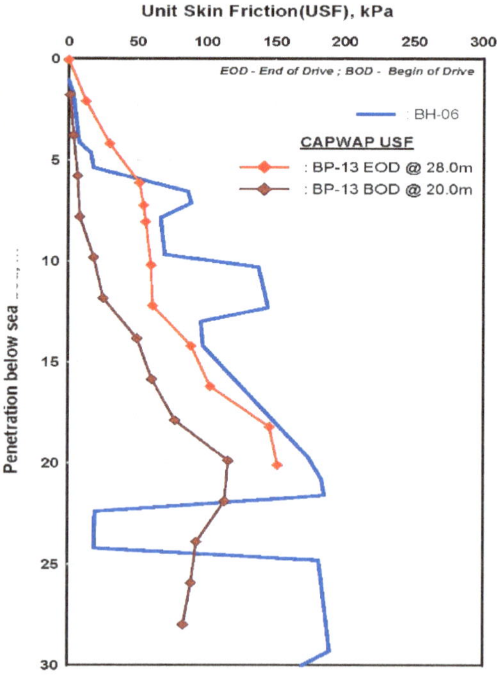

Figure 6.7.2-9 Unit Skin Friction (USF) vs. Pile penetration -Setup (SBD, 2m OD Monopiles)

Further estimated USF from EOD CAPWAP at final penetration is compared with the estimated USF from BOD CAPWAP (20m for SBD pile BP13 and 32m for LBD of BP1 & BP2) to establish the fact that soils in the vibro driven depths have exhibited notable setup. The **Figure 6.7.2-9** and **Figure 6.7.2-10** show the USF profiles respectively.

The lateral pile analysis is carried for the BP1 pile because it was driven to a deeper penetration of 22.5m using vibro hammer. The back-calculated as-installed shear strength parameters for BP1 monopile are summarized in **Table 6.7.2-9.** As shown in **Figure 6.7.2-10 and Figure 6.7.2-10,** the USF values of EOD ranged from 1/3rd to ½ of the static USF values, and the USF values (BOD) after the setup are in close agreement with the static USF values. This reduction in soil strength due to monopiles driving (using vibratory + impact hammer) and soil set up with the time is in line with the expectation. No unusual or severe soil strength reduction due to Vibratory driving is observed

PS: The USF from BOD and EOD CAPWAP for BP3 and BP 4 are not presented here.

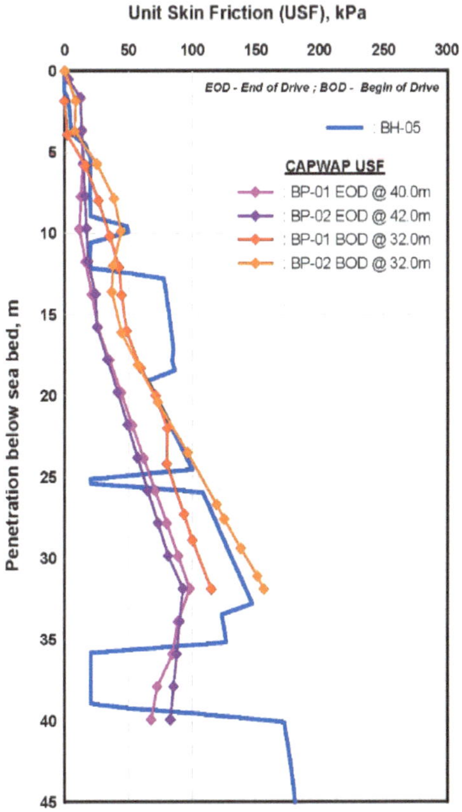

Figure 6.7.2-10: Unit Skin Friction (USF) vs. Pile penetration -Setup (LBD. 4m OD Monopiles)

Soil type	Top Depth, m	Bottom depth, m	Unit weight, kN/m³	phi (deg)	Shear strength, Su (kPa)	Modulus of Elasticity, E (kPa)
sand	0	1.7	8.1	40.00	-	40000
sand	1.7	3.72	8.55	33.28	-	30000
sand	3.72	5.73	9.35	24.60	-	30000
sand	5.73	7.74	10.15	20.00	-	25000
sand	7.74	9.75	10.15	20.00	-	25000
clay	9.75	10.25	9.9	-	51	10131
sand	10.75	11.76	9.4	20.00	-	19500
clay	11.76	13.77	9.95	-	66	13138
clay	13.77	15.79	9.95	-	81	16100
clay	15.79	17.8	9.95	-	109	21738
clay	17.8	19.81	9.85	-	137	27375
clay	19.81	21.82	9.95	-	165	33006
clay	21.82	22.75	9.95	-	193	38644
clay	22.75	23.83	9.95	-	193	38644
clay	23.83	25.84	9.75	-	221	44281
clay	25.84	27.86	10	-	250	49913
clay	27.86	29.87	10	-	278	55550
clay	29.87	31.88	10	-	306	61188
clay	31.88	33.89	10	-	277	55431
clay	33.89	35.9	9.45	-	264	52750
sand	35.9	37.91	10.4	20.00	-	30000
sand	37.91	39.92	10.4	20.00	-	30000

Table 6.7.2-9: Soil shear strength parameters (As-installed)

Pile-Soil Interaction Study

Although pile monitoring data indicated soil resistance from the penetration depths where the vibratory hammer was used, and the compression capacity achieved exceeded the design compression capacity, the oil field operator wanted to ensure the lateral pile capacity and fixity depth were not negatively affected by the vibratory driving. The as-installed soil shear strength parameters summarized in **Table 6.7.2-9** were used for modeling the Pile Soil Interaction (PSI) in Plaxis-3D software.

The following case was analyzed to study the effect of vibratory + impact hammer driving on lateral pile capacity for BP1 pile (which was installed to deeper depth using vibrohammer).

Case: As-installed soil strength parameters derived from EOD CAPWAP USF and applying setup of 1.59 only for penetrations where the impact hammer is used.

The soil and pile model in Plaxis-3D are shown in **Figure 6.7.2-11** and **Figure 6.7.2-12**. The tomography of total lateral displacements and fixity depths are shown in **Figure 6.7.2-13** and the values are summarized in **Table 6.7.2.10.**

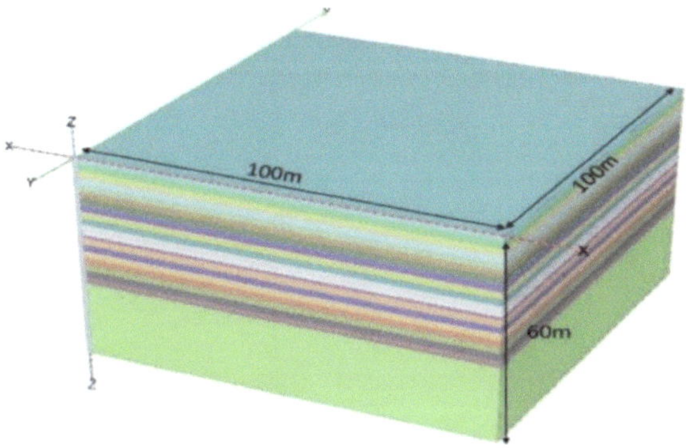

Figure 6.7.2-11 : Soil model boundaries in Plaxis-3D

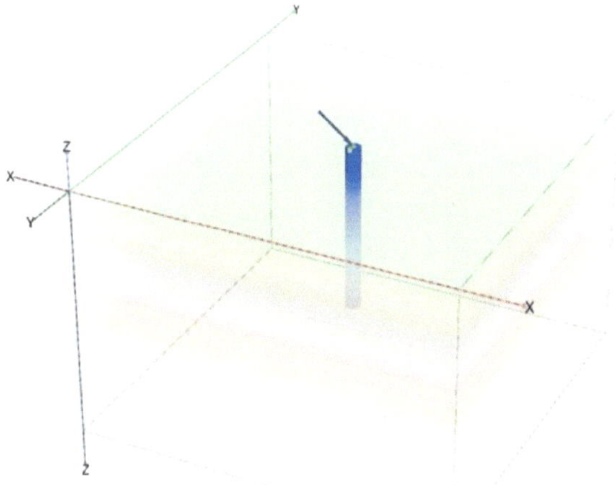

Figure 6.7.2-12 : Soil model boundaries in Plaxis-3D

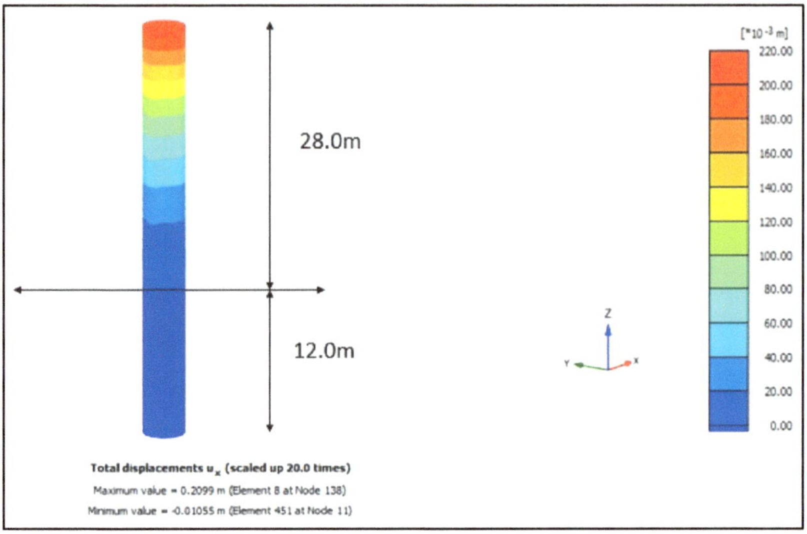

Figure 6.7.2-13 : Lateral displacement and Fixity, LBD

Monopile	Plaxis-3D (Current study)	
	Max. Lateral displacement, mm	**Fixity, m**
LBD, BP1	180	28

Table 6.7.2-10: Summary of total lateral displacement and fixities for LBD monopiles

Conclusions

The apprehension of the severe reduction in soil strength due to vibratory pile driving compared to impact pile driving was studied. The as-installed soil strength parameters, which account for vibratory and impact hammer driving, were derived from CAPWAP estimated USF and used for modeling the pile-soil interaction in Plaxis-3D. The study revealed that the final monopile penetrations achieved were suitable for lateral pile capacity & fixity and that there was no negative effect on soil strength degradation due to vibratory pile driving. The pile top displacement estimated by Plaxis 3D was less than 10% of the pile diameter, which was the specification and the fixity depth as per Plaxis 3D was shallower than the installed depth. The above results are an illustration for LBD pile only. Similar analyses were performed on SBD piles in Plaxis 3D and found to satisfy the lateral capacity requirement.

The Contractor analyzed in SACS for unity checks and confirmed that the structural requirements were also met.

6.7.3 Case Study-3 (Dr Ramesh Gangisetti, Ms Anusha Bekal & Mr Srinivas Nagappa)

Problem Statement

Pile driving monitoring is required to accurately estimate the available pile capacity during driving. However, if pile monitoring data is not available then pile driving data viz., blow counts and energy readings from the hammer console can be used to simulate the observed driving conditions in GRLWEAP software and estimate the pile capacity. This case study deals with the hindcast analysis for estimating the pile capacity for pipe piles installed at an offshore field in Saudi Arabia.

Project Details

A four-leg jacket was found on a pile foundation consisting of 1.067m OD pipe piles with varying wall thicknesses from 38.1mm to 44.45m. The final penetrations to which the piles were driven using hydraulic hammers IHC-S280, IHC-S600, and IHC-S900 hammers varied from 25m to 37m. The piles were initially driven to 18m using IHC-S80 and thereafter to their respective final penetrations using IHC-S600 and IHC-S900 hammers. The jacket plan is shown in **Figure 6.7.3-1.**

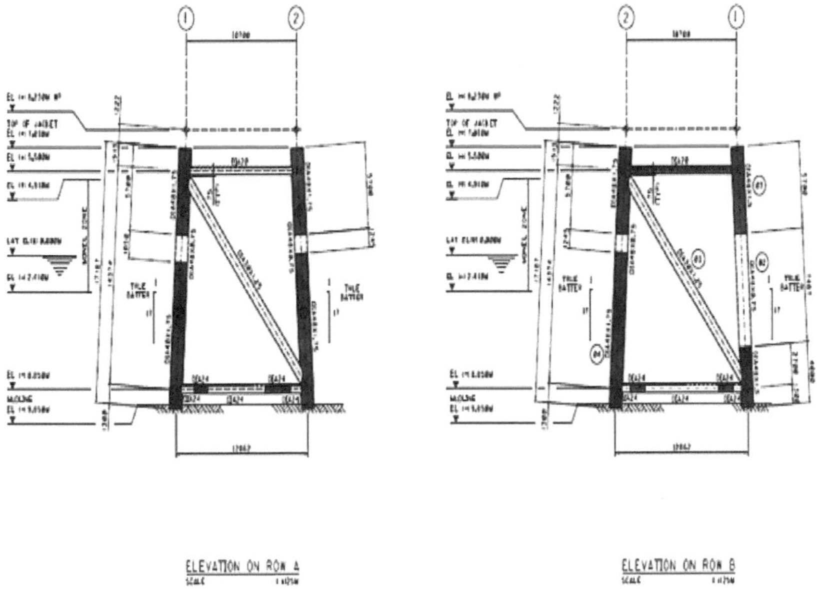

Figure 6.7.3-1: Jacket Plan

Soil Conditions

The site-specific geotechnical borehole indicated that the soil strata consist predominantly of dense to very dense Silty SAND until the final pile penetration depth of 37m with a 2m thick Hard slightly cemented CLAY at 16.9m. The sands are locally cemented in the layers at 7.9m and 18.9m. The soil profile is shown in **Table 6.7.3-1.**

Unit	Depth below Seafloor (m)		Soil Description	Strata thickness (m)
	From	To		
1	0	0.85	Very dense silty fine to medium siliceous carbonate SAND	0.85
2	0.85	2.1	Medium dense to dense silty fine to medium siliceous carbonate SAND	1.25
3	2.1	2.8	Very dense slightly silty fine to medium siliceous carbonate SAND	0.7
4	2.8	3.5	Medium dense slightly silty fine to medium Silica SAND	0.7
5	3.5	6.6	Medium dense to dense silty fine to medium calcareous silica SAND	3.1
6	6.6	7.9	Dense slightly silty fine to medium Silica SAND	1.3
7	7.9	9.35	Very dense locally cemented very silty calcareous silica SAND	1.45
8	9.35	11.1	Dense to very dense locally cemented silty silica SAND	1.75
9	11.1	16.9	Very dense silty fine to medium silica SAND	5.8
10	16.9	18.9	Hard slightly cemented CLAY	2
11	18.9	20.5	Dense to very dense locally cemented silty fine silica SAND	1.6
12	20.5	35.1	Dense to very dense silty fine silica SAND	14.6
13	35.1	46.7	Very dense silty fine to medium silica SAND	11.6

Table 6.7.3-1: Soil stratigraphy at the Jacket location

Pile Capacity Assessment

The objective is to simulate the observed driving condition in GRL WEAP software and back-calculate the pile capacity. The hammer-pile-soil system is modeled, and the hammer efficiency is adjusted until the computed energy (ENTHRU) is matched with the energy readings from the hammer console. Then for a given pile capacity and a pile penetration, blow count is computed. The pile capacity values are iteratively adjusted until the computed blow counts match the observed field blow counts. This procedure is iterated for all pile penetrations. The **Table 6.7.3-2** shows the USF, UEB, Soil quake, and damping values considered as the starting point of the analysis.

Soil Type	Depth	Unit Shaft Resistance	Unit Toe Resistance	Quake		Damping	
				Shaft	Toe	Shaft	Toe
	(m)	(kPa)	(kPa)	(mm)	(mm)	(sec/m)	(sec/m)
Sand	0	0	0	2.54	2.54	0.164	0.5
Sand	0.85	3.51	380.38	2.54	2.54	0.164	0.5
Sand	0.85	2.84	228.23	2.54	2.54	0.164	0.5
Sand	2.1	7.43	597.6	2.54	2.54	0.164	0.5
Sand	2.1	9.2	996	2.54	2.54	0.164	0.5
Sand	2.8	11.98	1297	2.54	2.54	0.164	0.5
Sand	2.8	9.68	518.8	2.54	2.54	0.164	0.5
Sand	3.5	12.5	670	2.54	2.54	0.164	0.5
Sand	3.5	15.47	1005	2.54	2.54	0.164	0.5
Sand	6.6	32.23	2093.1	2.54	2.54	0.164	0.5
Sand	6.6	32.23	2790.8	2.54	2.54	0.164	0.5
Sand	7.9	39.13	3388.8	2.54	2.54	0.164	0.5
Sand	7.9	47.46	4236	2.54	2.54	0.164	0.5
Sand	9.35	57.2	5106	2.54	2.54	0.164	0.5
Sand	9.35	57.2	4595.4	2.54	2.54	0.164	0.5
Sand	11.1	68.09	5469.53	2.54	2.54	0.164	0.5
Sand	11.1	68.09	8508.15	2.54	2.54	0.164	0.5
Sand	16.9	102.5	12811.75	2.54	2.54	0.164	0.5
Clay	16.9	144	3015	2.54	2.54	0.164	0.5
Clay	16.9	147.7	3015	2.54	2.54	0.164	0.5
Sand	18.9	105.5	9118.13	2.54	2.54	0.164	0.5
Sand	18.9	105.5	9917.33	2.54	2.54	0.164	0.5
Sand	20.5	123.5	14400	2.54	2.54	0.164	0.5
Sand	26.8	162.3	14400	2.54	2.54	0.164	0.5
Sand	26.8	162.3	12950	2.54	2.54	0.164	0.5
Sand	35.1	175	12950	2.54	2.54	0.164	0.5
Sand	35.1	200	14400	2.54	2.54	0.164	0.5
Sand	39.1	200	14400	2.54	2.54	0.164	0.5

Table 6.7.3-2 : Soil resistance input for wave equation analysis

The shaft Gain/Loss factor was kept constant at 1.0 and the static USF and UEB were adjusted to match blow count at penetrations where local increase or decrease in blow count was observed. The final shaft and toe Gain/Loss factors and the dynamic parameters derived are summarized in **Table 6.7.3-3.**

Pile	Penetration . m	Gain/ loss factor		Damping, sec/m		Quake, mm	
		Shaft	Toe	Shaft (sand/clay)	Toe	Shaft	Toe (sand/clay)
A1P1	0-6.8	1	0.25	0.164/0.656	0.5	2.54	8.89/17.78
	6.8-37.3	1	1	0.164/0.656	0.5	2.54	8.89/17.78
A2P1	0-0.7	1	0.25	0.164/0.656	0.5	2.54	8.89/17.78
	7.0-37.1	1	1	0.164/0.656	0.5	2.54	8.89/17.78
A2P1Restrike	37.1-37.5	1	1	0.164/0.656	0.5	2.54	8.89/17.78
B1P1	2.0-7.0	1	0.5	0.164/0.656	0.5	2.54	2.54/2.54
	7.0-25.0	1	1	0.164/0.656	0.5	2.54	2.54/2.54
B2P1	2.0-7.0	1	0.5	0.164/0.656	0.5	2.54	2.54/2.54
	7.0-25.0	1	1	0.164/0.656	0.5	2.54	2.54/2.54

Table 6.7.3-3: Summary of shaft and toe Gain/ Loss factors and dynamic parameters

The **Figure 6.7.3-2, Figure 6.7.3-3, Figure 6.7.3-4 and Figure 6.7.3-5** show the observed Energy/Blow count vs computed values match for piles A1, A2, B1, and B2 respectively.

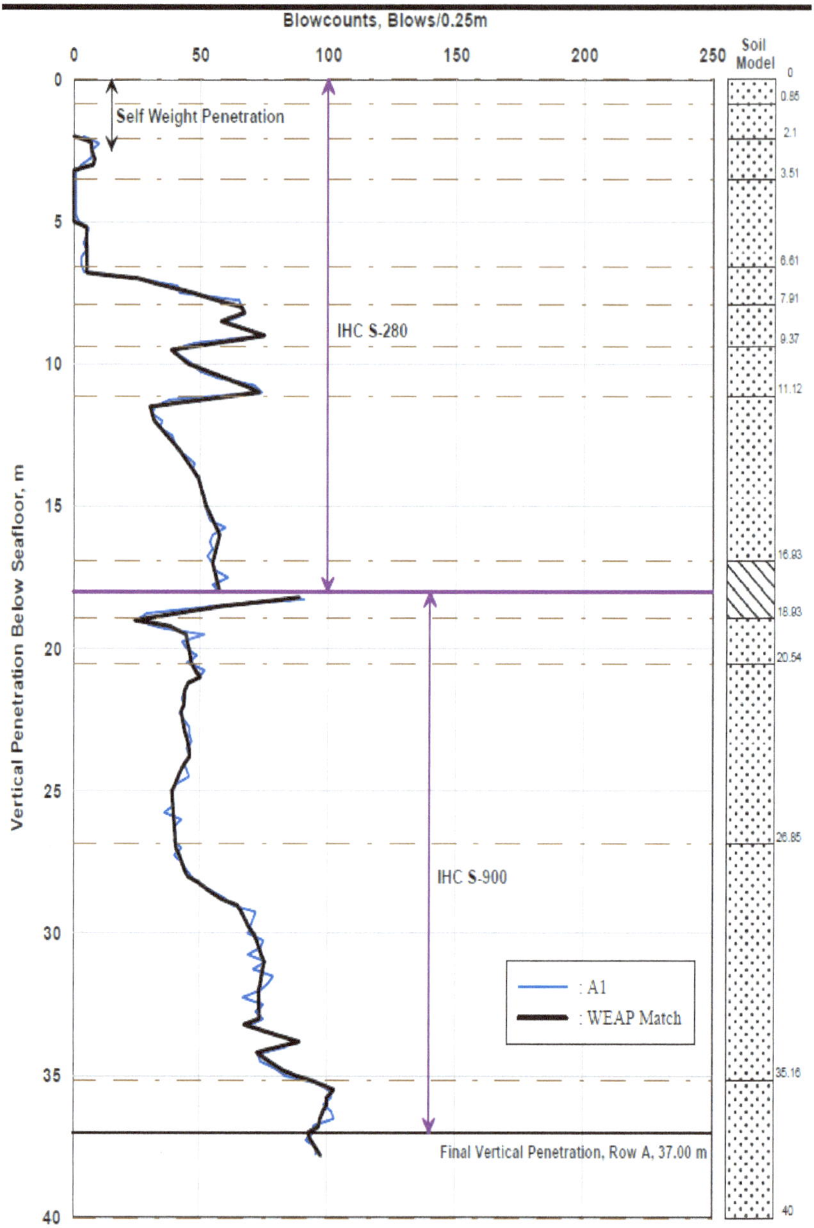

Figure 6.7.3-2: Observed vs Weap computed blow counts (Pile A1)

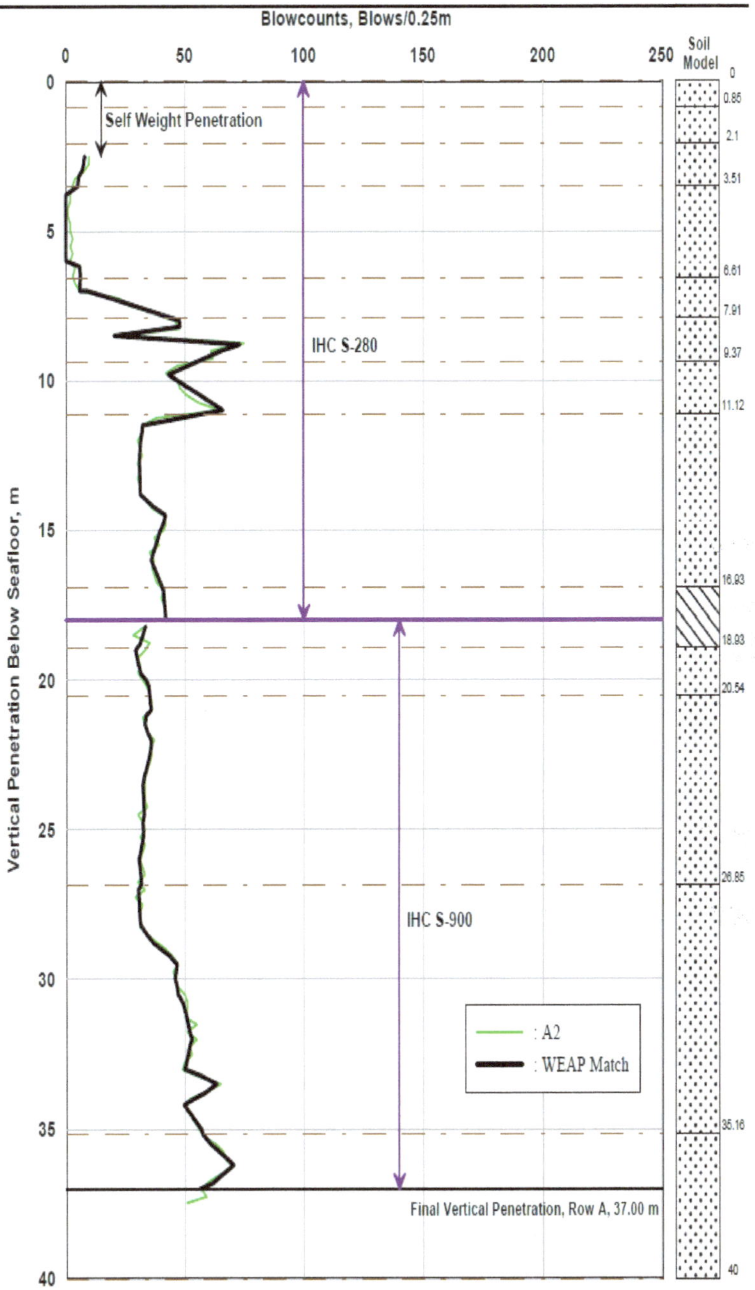

Figure 6.7.3-3: **Observed vs Weap computed blow counts (Pile A2)**

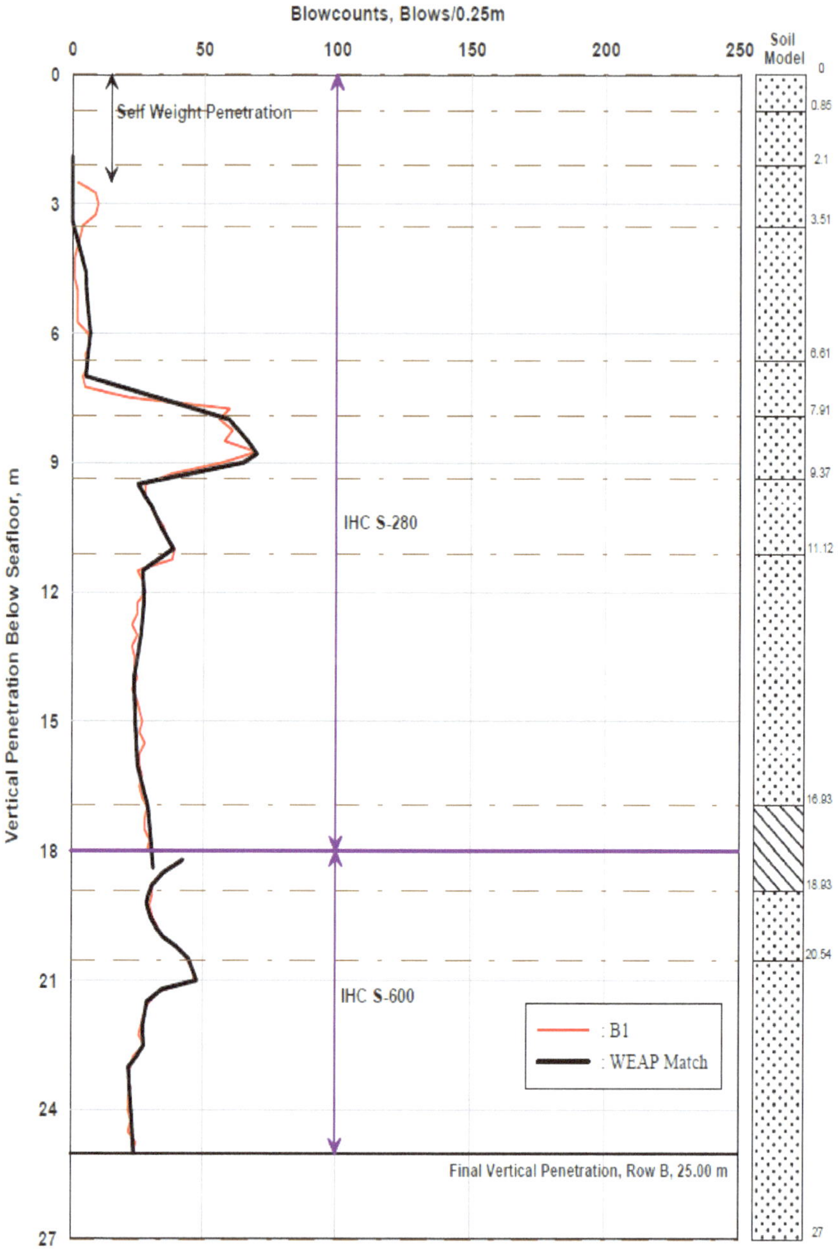

Figure 6.7.3-4: **Observed vs Weap computed blow counts (Pile B1)**

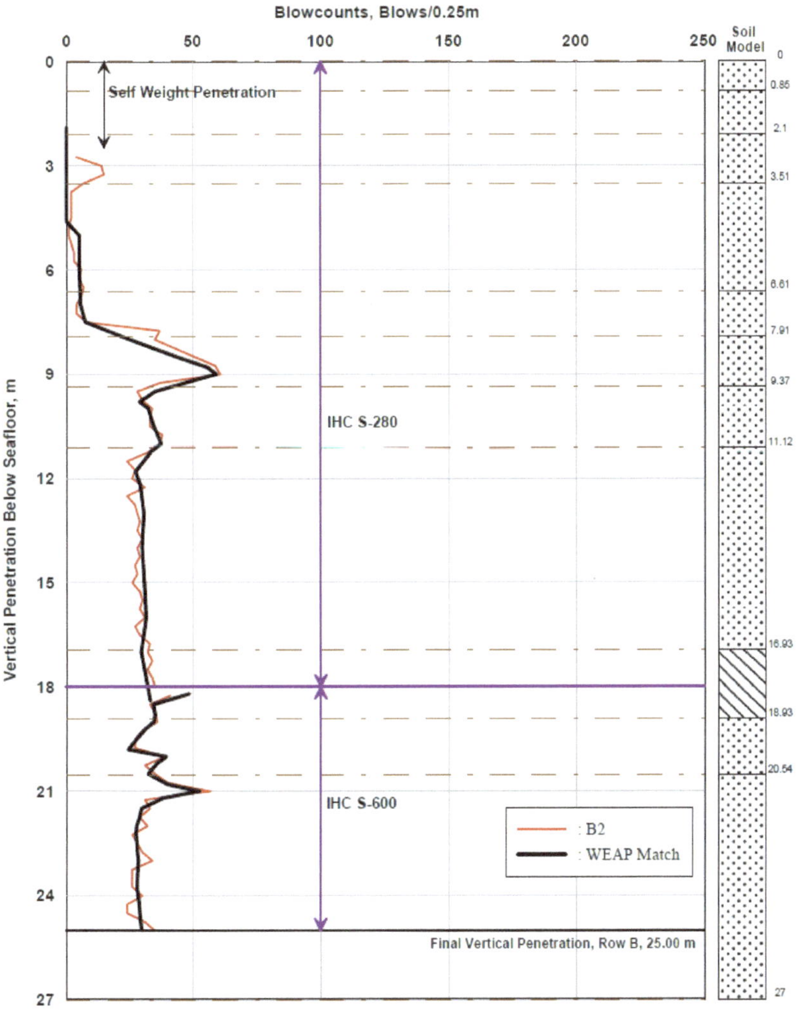

Figure 6.7.3-5: **Observed vs Weap computed blow counts (Pile B2)**

A bearing graph analysis was done separately to match the restrike blow counts and consequently obtain the restrike compression capacity as shown in **Figure 6.7.3-6.**

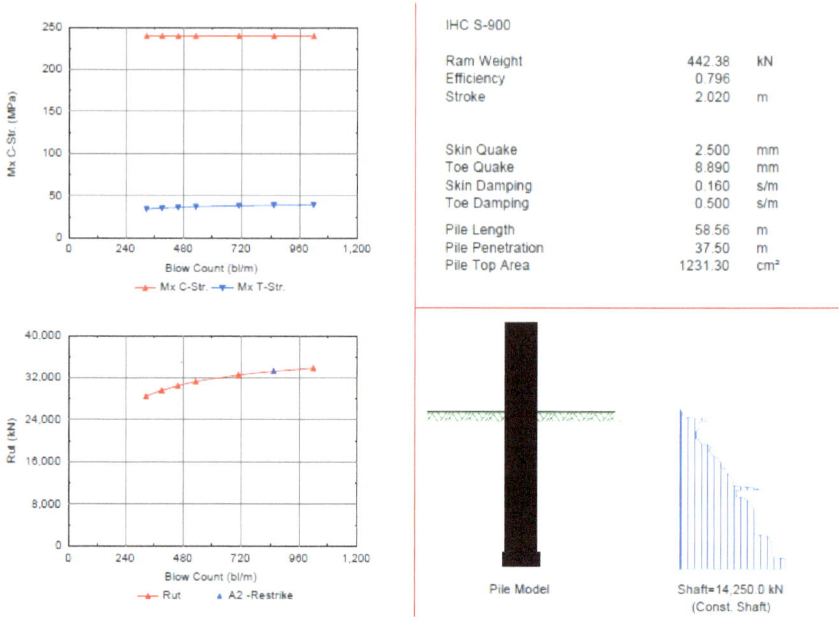

Figure 6.7.3-6: Bearing Graph

The back-calculated SRD was considered as the available pile capacity in compression. The back-calculated pile capacities at EOD and Restrike were used to calculate soil setup and the same was used for non-restrike piles to estimate the long-term static capacity. The tension capacity was calculated as 80% of the shaft compression capacity. The summary of pile capacities is summarized in **Table 6.7.3-4.**

Pile	Pile capacity (Back-calculated from GRLWEAP)	
	Compression, MN	Tension, MN
A1	29.9	12.5
A2	32.1	11.4
B1	18.8	5.7
B2	20.5	6.1

Table 6.7.3-4: Summary of back-calculated compression and tension pile capacities at final penetration

Conclusions

In the event pile monitoring data is unavailable, the observed pile driving behavior can be modeled in GRLWEAP software to back-calculate the available pile compression capacity during driving. The static USF and UEB as per the API RP2 Geo, along with the standard soil dynamic parameters outlined in the GRLWEAP manual shall be the starting point of the iterative procedure. These values will be adjusted until the field blow counts are matched, at which point the soil resistance to driving will represent the available pile compression capacity.

6.4.7Associated Risks and Engineering Solutions

In the vast offshore area of Saudi Arabia, there is a wide spectrum of geotechnical risks associated with the offshore engineering works.

In fields such as Marjan, the stratigraphy is made of heavily over-consolidated cohesive deposits and very scarce presence of cemented layers. The over-consolidation ratio of these deposits in the shallow depths of 20 to 30m lies in the range of 15 to 25 and in the range of 3 to 6 at deeper depths. The shear strength of such deposits is significantly high ($\geq$ 250kPa) as compared to the stratigraphy found in South-East Asia (generally Su is <200 kPa).

In the other fields of offshore Saudi Arabia, the cemented gypsum layers are of very common occurrence. Therefore, cases of pile refusal are inevitable. A thorough geotechnical investigation should be conducted to predict the occurrence of refusal driving and engineering solutions are recommended to be planned prior to mobilization of the installation vessels. If at refusal, the evaluated pile capacity, as installed, has achieved the required pile capacity, then the pile is accepted as is after checking on the lateral load capacity and bending moment. These are usually satisfied in most cases where the refusal occurs at depths deeper than say 30m from the mudline. If the required capacity is not achieved at refusal, a variety of remedial solutions are available for refusal cases. Some of which are discussed below,

- If the thickness of the cemented layer is less, weakly cemented and no multiple cemented layers are expected further: Then, the cemented layers are reamed from the depth of refusal to the bottom of the layer, and the pile is then driven further.
- If the thickness of cemented layer is more, with multiple layers and strong to extremely strong in terms of strength:

An insert pile of a smaller diameter than the initial pile is prepared. The cemented layers are reamed through, and the insert pile is then lowered through the initial pile and driven to the target penetration. The annular area of the insert pile is then grouted.

- If the stratigraphy is predominantly cemented layers from mudline to deeper depths: The methodology of a drilled and grouted pile is adopted. This methodology of deep foundation installation is very common in the Abu Dhabi region **(see chapter 5).**

The mudline of offshore Saudi Arabia region: It is also common to encounter cemented layers at mudline, where the bearing capacity of the mud mat is evaluated with the conventional methods assigning sand parameters. In cases where the mudline consists of very soft to soft clay layers of greater thickness, in addition to the large mud mats, peripheral and/or internal skirts are provided to mitigate bearing capacity problem of the mud mats. The depth of the skirts is usually limited to about 2m. Further, numerical methods shall be used to evaluate settlement and tilt of the mud mats. In case of skirted mud mats, the lateral stability is verified by the undrained shear strength at the tip of the skirt (weak layer), while the bearing capacity is derived by the strong layer below the skirt tip due to the influence depth.

The Jack up rig leg penetrations expected in this region is shallower. However, occasionally, penetrations range from about 5 to 10m. In cases of well-cemented mudlines, the tip of the spud can hardly penetrate, in such a case, a pinned analysis shall be considered using finite element methods. Though the risk of punch-through is less in this region, it is recommended that accurate analysis must be made prior to deployment of the rigs.

The existing footprints from initially positioned Jakup Rigs are also an important factor to be considered as they may cause instability. This must be reviewed thoroughly from the geophysical reports on a case-by-case basis.

Subsea pipeline design starts with client-provided data on pipeline parameters, crucial for pipeline-soil interaction studies. Extensive shallow geotechnical investigations along the pipeline route aid in zoning and deriving ground models. Parameters for Pipeline Soil Interaction analysis, like embedment, lay factors, axial and lateral stiffness, effective stress, and shear modulus are determined based on pipeline and soil properties.

Proper soil characterization and zoning guarantee reasonably accurate PSI analysis, ensuring pipeline stability. Additionally, scour and liquefaction protection measures are designed considering pipeline embedment, further enhancing resilience. This meticulous approach integrates engineering principles and site-specific data to optimize pipeline performance and longevity, vital for the reliable transportation of fluids in challenging offshore environments.

6.8 Lessons from the Experienced

6.8.1 Leadership Through Experience: A 25-Year Journey in Global EPC Industry – Mr Marc Girodroux.

Synopsis

Marc Girodroux's journey from a fabrication yard to leadership positions exemplifies the value of diverse experience, international exposure, and people-centric leadership in the EPC industry. His insights into technical challenges, particularly in geotechnical aspects, and the importance of knowledge sharing provide valuable lessons for industry professionals. As the region enters what appears to be a growth cycle, his emphasis on developing both business and people presents an optimistic outlook for the industry's future.

Introduction

In an insightful conversation, Marc shares his remarkable 25-year journey in the Engineering, Procurement, and Construction (EPC) industry, from his early days at an Angolan fabrication yard to his current leadership position in Dubai. This interview explores the crucial lessons learnt, challenges overcome, and the importance of people-centric leadership in the global EPC sector.

Interviewee Profile

Marc is a seasoned industry leader with over two decades of experience in the EPC sector. Starting his career in fabrication yards in Angola, he has progressively held positions across various domains including sales, marketing, general management, engineering, and supply chain. Recently relocated to Dubai, Marc brings extensive international experience and a strong focus on collaborative leadership. His expertise spans deep-water works, offshore installations, and complex project management.

The Talk

CRP: Dr CR Parthasarathy

MG: Marc Girodroux, Vice President, Middle East, Subsea 7

CRP: Could you share your career journey and how you reached this leadership level?

MG: I started my career about 25 years ago, beginning at a fabrication yard in South Angola working on jackets for the offshore industry. Over time, I've gained experience across various areas including fabrication, sales and marketing, general management, engineering, and supply chain. I recently moved to Dubai about two months ago.

CRP: What significant experiences have shaped your career?

MG: Two major factors have influenced my career path. First, working abroad from a young age has shaped my thinking and approach to interacting with people. I would call that an eye opener, I would say, making you more conscious about what's happening around. Second, I've actively pursued diverse roles and expertise areas, which have created numerous learning and development opportunities.

CRP: What principles and practices do you follow to stay ahead in the industry?

MG: People are central to our success. You can't achieve everything alone - it's about working collaboratively with others to reach targets. This principle has been fundamental to our planning and will continue to be crucial in the coming months.

CRP: Could you share any significant geotechnical challenges from your projects?

MG: One notable experience was during our initial deep-water works in Angola, particularly with suction piles. We learned the critical importance of soil conditions and the friction between soil and pile, which significantly affects the strength and performance of installations.

CRP: How important is ground characterization for EPC projects?

MG: It's crucial. Unknown soil conditions can transform an easy installation into a complex one. Within our company, we've faced significant challenges due to insufficient knowledge of soil conditions and geotechnical aspects, which can have substantial impacts on project execution.

CRP: What advice would you give to the next generation of industry professionals?

MG: Sharing experiences and lessons learned is vital. We often see cycles where mistakes are repeated because knowledge isn't properly recorded or maintained. Documenting these experiences helps future professionals avoid repeating the same mistakes.

CRP: Who has inspired you as a role model in your career?

MG: Rather than one specific person, I've been influenced by multiple leaders' management styles and their approaches to collaboration. However, the person who hired me 25 years ago had a unique style focused on trusting and empowering people, which left a lasting impression.

CRP: What's your vision for the future?

MG: My immediate focus is on properly managing and developing in this region. The goal is to develop profitable business while growing our people both individually and collectively.

CRP: What's your perspective on the industry outlook for the next 3-5 years?

MG: The region shows tremendous potential with numerous developments planned. After challenging periods including COVID and the mid-2015 oil crisis, we're seeing signs of a long, positive cycle with many projects and developments in the pipeline.

CRP: How do you maintain work-life balance?

MG: While admitting it's challenging, I focus on inspiring my children and sharing my passion for the industry. My son, influenced by early exposure to fabrication yards, is now pursuing engineering, which is particularly rewarding.

PS: Currently, Subsea 7 and Saipem have merged in to Saipem 7

7

TEN PRINCIPLES FOR LIVING A PURPOSEFUL AND FULFILLING LIFE

7.1 Chapter Overview

In this transformative chapter based on authors own personal experience, explore a holistic approach to personal growth, success, and well-being through ten fundamental principles that can reshape one's life trajectory.

1. Dream Big, But Plan Strategically

Dreams are the fuel of human potential, but they must be accompanied by actionable strategies. While having ambitious goals is crucial, breaking them down into smaller, achievable milestones is the key to sustainable progress. Consider your career over a 3-5 year horizon but simultaneously create short-term goals that can be accomplished within weeks or months.

2. Start Small, Build Consistently

The secret to meaningful change lies in starting small and maintaining discipline. Take health as an example: instead of overwhelming yourself with an aggressive fitness regime, beginning with just 10 minutes of daily physical activity. Consistency is more important than intensity. Discipline beats motivation.

By dedicating just 10 minutes daily and gradually increasing your commitment, you transform sporadic efforts into a sustainable lifestyle.

3. Financial Discipline: Live Below Your Means

Financial health is not about earning more but managing what you have wisely. The principle is simple yet powerful: live below your means. Start by saving say just 5% of your earnings through mindful choices like carpooling, sharing resources, and reducing unnecessary expenses etc., This approach, inspired by the Japanese concept of Kaizen, emphasizes incremental improvement. Further, equip yourself with financial literacy to earn financial freedom.

4. Continuous Learning and Skill Development

Skill acquisition is not about massive time investments but consistent, focused effort. Dedicate 10-15 minutes daily to researching and practicing a skill you want to develop. By the end of a month, you'll have accumulated significant knowledge, and within 3-4 months, you would have developed a substantial new competency. Learning new things is a lifelong journey.

5. Cultivate the Right Environment (Context Setting)

Your progress is significantly influenced by your surroundings (People and atmosphere). Seek out like-minded individuals who support, inspire, challenge, impress and hold you accountable. The people you interact with most frequently can either accelerate or hinder your personal growth. Set the Context wisely.

6. Develop a Personal Code of Honor (to uphold value)

Create a set of personal rules that uphold your core values. For instance, in health as a value, example of the rules can be

- Not consuming toxic substances daily (be mindful of eating)
- Ensuring 7-8 hours of sleep
- Drinking adequate water
- Engaging in at least 30 minutes of physical activity daily

(Other values: Loyalty, Integrity, Honesty, Compassion, Mindfulness etc).

7. Embrace Life's Dualities

Recognize that life consists of both positive and negative experiences. Like day and night, good and bad are two sides of the same coin. When faced with challenges, maintain a balanced perspective and focus on doing your duty with sincerity and honesty. Embrace and accept equally, High & Low, Success & Failure, Gain & Loss etc.,

8. Develop a Growth Mindset

View every experience as an opportunity to learn and grow. Whether you succeed or face setbacks, approach each situation with curiosity and resilience.

9. Practice Mindfulness and Intentionality

Be conscious of how you spend your time. Avoid mindless scrolling and social media consumption (experience digital sunset). Instead, invest your time in activities that contribute to your personal and professional development.

10. Appreciate the Gift of Existence and Celebrate success

Recognize the miracle of human existence. We have the unique ability to dream, grow, and transform. Every moment is an opportunity to live fully and purposefully. Celebrate all successes, small or big doesn't matter.

7.2 Conclusion

These principles are not rigid rules, but flexible guidelines designed to help you navigate life's complexities. The journey of personal growth is ongoing, requiring patience, consistency, and a commitment to continuous improvement. Remember, transformation happens not through grand gestures, but through small, consistent actions performed with intention and heart.

7.3 Reflection Questions

- What small step can you take today towards your larger goals?
- How can you create a more supportive environment for your growth?
- What personal code of honor will you develop?

PS: This chapter is a prelude to my next book on Ultimate personality development.

LIFE AUDIT & GOAL SETTING

Welcome to the Life Audit section, the first & arguably the most important step on your self-improvement journey. This section serves as your starting point, helping you to discern the areas of your life that you want to focus on and work towards achieving through self-help or through mentoring or through a coach.

Understanding the domains that require attention is key to setting actionable and impactful goals. The Life Audit aims to provide you with clarity & perspective which allows you to set meaningful and compelling goals.

Use the wheel of life in the next page to identify how satisfied you are with each area of your life. Rate each segment on a scale of 1 to 10, with 1 being "Needs Significant Improvement" and 10 being "Completely Satisfied". Be honest with your self-assessment to gain the most benefit from this exercise.

When you're done, move on to the questions on the following page.

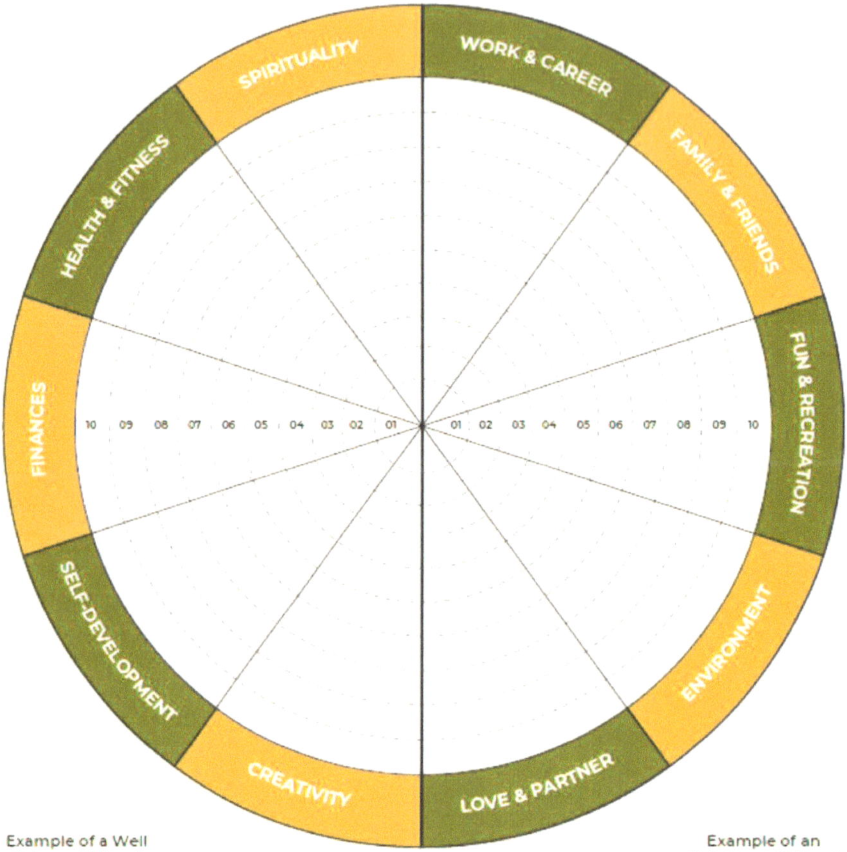

Example of a Well Balanced Wheel

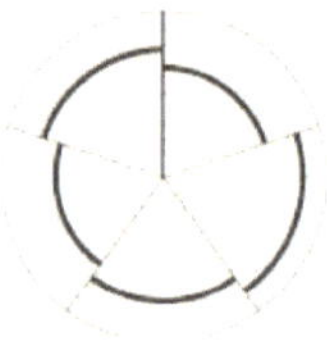

PS! The "Environment" means your physical location, your work environment, home environment, etc

Example of an Unbalanced Wheel

Use the questions below to reflect on your results

1. What areas of your life do you feel that you need to work most on? Why?

2. If you focused on just one area of your life right now, which one would have the highest positive impact on your life?

3. Write down your goals (max 3) in that area of your life (time bound)

Be Amazing and Awesome

BIBLIOGRAPHY

1. Abbs, A. F. "Lateral pile analysis in weak carbonate rocks." In *Geotechnical practice in offshore engineering*, pp. 546-556. ASCE, 1983.
2. Abbs, A. F., and A. D. Needham. "Grouted piles in weak carbonate rocks." In *Offshore Technology Conference*, pp. OTC-4852. OTC, 1985.
3. Al-Dabbagh, Mohammad E. "The Arabian plate: unique fit of the earth's surface jig saw puzzle." Arabian sJournal of Geosciences 7 (2014): 3297-3307.
4. American Petroleum Institute (API) RP 2GEO, 1st Edition Apr 2011, Geotechnical and Foundation Design considerations.
5. Anderson and Lunne (1994). "A method to determine conductor setting depth in clay". Vol 1
6. Angemeer, James, Emery G. Carlson, and James H. Klick. "Techniques and results of offshore pile load testing in calcareous soils." In *Offshore Technology Conference*, pp. OTC-1894. OTC, 1973.
7. Ashford, S. A., and Rollins, K. M. (2002). TILT: The Treasure Island Liquefaction Test: Final Report, Report SSRP-2001/17, Department of Structural Engineering, University of California, San Diego.
8. Bashir, Yasir, Deva Prasad Ghosh, Hammad Tariq Janjuhah, and Chow Weng Sum. 2018. "Diffraction Enhancement Through Pre-Image Processing: Applications to Field Data, Sarawak Basin, East Malaysia" *Geosciences* 8, no. 2: 74. https://doi.org/10.3390/geosciences8020074
9. Bastia, Rabi., Radhakrishna, M. Basin Evolution and Petroleum Prospectivity of the Continental Margins of India. Netherlands: Elsevier, 2012.
10. Bowles, Joseph E., and Yingzhong Guo. *Foundation analysis and design.* Vol. 5. New York: McGraw-hill, 1996.
11. Boulanger, Ross W., and I. M. Idriss. "Evaluation of cyclic softening in silts and clays." *Journal of geotechnical and geoenvironmental engineering* 133, no. 6 (2007): 641-652.

12. Bullock, Paul J., John H. Schmertmann, Michael C. McVay, and Frank C. Townsend. "Side shear setup. I: Test piles driven in Florida." *Journal of Geotechnical and Geoenvironmental Engineering* 131, no. 3 (2005): 292-300.
13. CLAROM, Stability and Operation of Jackups. France: Editions Technip, 1993.
14. Deshmukh, Soumen, Anshumita Ganguly, Imtimenla Ao, Kalingarajah Karvannan, Radha Krishan, and A. Sinha. "Syn-Rift Sequence: A Major Play for Future Exploration, KG Basin." *ONGC Bulletin* 52, no. 2 (2017): 61–74.
15. Derakhshani, Reza, and Godratollah Farhoudi. "Existence of the Oman Line in the Empty Quarter of Saudi Arabia and Its Continuation in the Red Sea." *Journal of Applied Sciences* 5 (2005): 745–752. https://doi.org/10.3923/jas.2005.745.752.
16. Directorate General of Hydrocarbons, Government of India. *Mumbai Offshore Basin.* https://www.dghindia.gov.in/assets/downloads/56cfd7ae55eb2Mumbai_Offshore_basin.pdf.
17. Directorate General of Hydrocarbons, Government of India. *The Kerala-Konkan Basin.* Accessed [9th July 2025]. https://dghindia.gov.in//assets/downloads/56cfd920029d9The_Kerala-Konkan_basin.pdf.
18. Directorate General of Hydrocarbons. *India's Hydrocarbon Outlook 2023–24*. New Delhi: Ministry of Petroleum & Natural Gas, Government of India, 2024. https://www.dghindia.gov.in/assets/downloads/ar/2023-24/136/.
19. Directorate General of Hydrocarbons. *Saurashtra Basin: Basic Introduction.* New Delhi: Government of India. Accessed July 9, 2025. https://dghindia.gov.in/assets/downloads/56ceb7d978a93Saurashtra_Basin_20.pdf.
20. DNVGL, DNV-RP-F114 Pipe-soil interaction for submarine pipelines, no. May. 2019.
21. Edgell, H. S. "Regional Stratigraphic Relationships of Arabia in Exploration for Oil and Gas. In, Short Course on Hydrocarbon Exploration." (1987): 1-44.
22. Franke, Kevin W., and Kyle M. Rollins. "Simplified hybrid p-y spring model for liquefied soils." *Journal of geotechnical and geoenvironmental engineering* 139, no. 4 (2013): 564-576.

23. Grötsch, Jürgen, Omar Suwaina, Ghiath Ajlani, Ahmed Taher, Reyad El-Khassawneh, Stephen Lokier, Gordon Coy, Erik van der Weerd, Shehadeh Masalmeh, and Johan van Dorp. "The Arab Formation in central Abu Dhabi: 3-D reservoir architecture and static and dynamic modeling." GeoArabia 8, no. 1 (2003): 47-86.
24. Hesse, S., S. Back, and D. Franke. "The Structural Evolution of Folds in a Deepwater Fold and Thrust Belt – A Case Study from the Sabah Continental Margin Offshore NW Borneo, SE Asia." *Marine and Petroleum Geology* 27, no. 2 (February 2010): 442–454. https://doi.org/10.1016/j.marpetgeo.2009.09.004.
25. Igoe, David, Giovanni Spagnoli, Paul Doherty, and Leonhard Weixler. "Design of a novel drilled-and-grouted pile in sand for offshore oil&gas structures." *Marine Structures* 39 (2014): 39-49.
26. International Trade Administration. 2021. "Energy Resource Guide – Indonesia – Oil and Gas." U.S. Department of Commerce. Accessed July 10, 2025. https://www.trade.gov/energy-resource-guide-indonesia-oil-and-gas?anchor=content-node-t7-field-lp-region-2-
27. International Trade Administration. 2023. "Energy Resource Guide – United Arab Emirates – Oil and Gas." U.S. Department of Commerce. Published November 25, 2023. Accessed July 10, 2025. https://www.trade.gov/country-commercial-guides/united-arab-emirates-oil-and-gas
28. ISO 3421:2022 - Petroleum and natural gas industries - Drilling and production equipment - Offshore conductor design, setting depth and installation
29. ISO 19905-1:2023 Oil and gas industries including lower carbon energy — Site-specific assessment of mobile offshore units-Part 1: Jack-ups: elevated at a site
30. International Society for Soil Mechanics and Geotechnical Engineering (ISSMGE), Geotechnical and Geophysical Investigations for Offshore and Nearshore Developments, September 2005.
31. International Standards Organisation (ISO), Petroleum and National Gas Industries – Specific Requirements for Offshore Structures – Part 8: Marine Soil Investigations, ISO 19901-8, 2014.
32. Jain, A. K., Dhiraj Mohan Banerjee, and Vivek S. Kale. "Tectonics of Western Margin of India." In *Tectonics of the Indian Subcontinent*, 525–555. Cham: Springer 2020. https://doi.org/10.1007/978-3-030-42845-7_9

33. Juirnarongrit, Teerawut, and Scott A. Ashford. "Soil-pile response to blast-induced lateral spreading. II: analysis and assessment of the p–y method." *Journal of Geotechnical and Geoenvironmental Engineering* 132, no. 2 (2006): 163-172.
34. Ledezma, Christian, and Jonathan D. Bray. "Probabilistic performance-based procedure to evaluate pile foundations at sites with liquefaction-induced lateral displacement." *Journal of geotechnical and geoenvironmental engineering* 136, no. 3 (2010): 464-476.
35. Liu, Lei, and Ricardo Dobry. "Effect of liquefaction on lateral response of piles by centrifuge model tests." (1995).
36. Madon, Mazlan. "Five decades of petroleum exploration and discovery in the Malay Basin (1968-2018) and remaining potential." (2021).
37. Madon, Mazlan. "Gas Hydrate Resource Potential of Deepwater Sabah, Malaysia: A Preliminary Assessment." *Bulletin of the Geological Society of Malaysia* 74 (November 2022): 1–15. https://doi.org/10.7186/bgsm74202201
38. Magoon, Leslie B., and Wallace G. Dow. "The petroleum system." (1994).
39. Matlock, Hudson. "Correlation for design of laterally loaded piles in soft clay." In *Offshore technology conference*, pp. OTC-1204. OTC, 1970.
40. Mesri, Gholamreza. "Discussion of "New design procedure for stability of soft clays"." Journal of the Geotechnical Engineering Division 101, no. 4 (1975): 409-412.
41. Ministry of Petroleum Resources, Deputy Ministry of Mineral Resources 1984. Geographic Map of the Arabian Peninsula, (showing oil field outlines). Map AP-5B-2; Scale 1:2,000,000; compiled by the United States Geological Survey and the Arabian American Oil Company.
42. Ministry of Statistics and Programme Implementation, Government of India. *Energy Statistics 2023: Chapter 1 – Reserves and Potential for Generation.* June2023. https://www.mospi.gov.in/sites/default/files/publication_reports/Energy_Statistics_2023/Chapter%201-Reserves%20and%20Potential%20for%20Generation.pdf.
43. Moh, Z. C. "Geotechnical engineering in Southeast Asia, past, present and future." *Geotechnical Engineering* 19, no. 1 (1987): 1-72.
44. Morad, Sadoon & Al-Aasm, Ihsan & Nader, Fadi & Ceriani, Andrea & Gasparrini, Marta & Mansurbeg, Howri. (2012). Impact of diagenesis on the spatial and temporal distribution of reservoir quality in the Jurassic

Arab D and C Members, offshore Abu Dhabi oilfield, United Arab Emirates. Geoarabia -Manama-. 17. 17-56. 10.2113/geoarabia170317.

45. Nauroy, J. F., and P. LeTirant. "Driven piles and drilled and grouted piles in calcareous sands." In *Offshore Technology Conference*, pp. OTC-4850. OTC, 1985.
46. Oil and Gas News (2023): "Lucky 7, Saudi Arabia's Well of Prosperity," Oil & Gas News, Aramco Review – History, September 2, 2023, Oil & Gas News, https://ognnews.com/Article/46303/Lucky_7,_Saudi_Arabia's_well_of_prosperity
47. Olson, Scott M., and Timothy D. Stark. "Liquefied strength ratio from liquefaction flow failure case histories." *Canadian Geotechnical Journal* 39, no. 3 (2002): 629-647.
48. Parthasarathy CR, Ramesh Gangisetti and Nikita Keshav (2016). "Case Studies – Predicting Soil Resistance to Driving in South East Asia", International Symposium on Frontiers in Offshore Geotechnics (ISFOG) 10-12 June 2015, Norway
49. Peng, Lee Chai, Mohd Shafeea Leman, Kamaludin Hassan, Bahari Md Nasib, and Rashidah Karim. Stratigraphic lexicon of Malaysia. Kuala Lumpur: Geological Society of Malaysia, 2004.
50. Powell, C. McA, S. R. Roots, and J. J. Veevers. "Pre-breakup continental extension in East Gondwanaland and the early opening of the eastern Indian Ocean." *Tectonophysics* 155, no. 1-4 (1988): 261-283.
51. PwC Indonesia. *Oil and Gas in Indonesia: Investment and Taxation Guide.* Jakarta: PricewaterhouseCoopers Indonesia, 2020. https://www.pwc.com/id/en/energy-utilities-mining/assets/oil-and-gas/oil-and-gas-guide-2020.pdf.
52. Ramprasad, T., P. Dewangan, M. V. Ramana, A. Mazumdar, S. M. Karisiddaiah, E. R. Ramya, and G. Sriram. "Evidence of Slumping/Sliding in Krishna–Godavari Offshore Basin Due to Gas/Fluid Movements." *Marine and Petroleum Geology* 28, no. 10 (2011): 1806–1816. https://doi.org/10.1016/j.marpetgeo.2011.02.007.
53. Riehle, James R., Frederic H. Wilson, Nora B. Shew, and Willis H. White. "Geology of Unga Island and the northwestern part of Popof Island: Chapter 2 in A geological and geophysical study of the gold-silver vein system of Unga Island, Southwestern Alaska." Open-File Report 99-136-2 (1999).

54. Rollins, Kyle M., J. Dusty Lane, and Travis M. Gerber. "Measured and computed lateral response of a pile group in sand." *Journal of Geotechnical and Geoenvironmental Engineering* 131, no. 1 (2005): 103-114.
55. Rollins, Kyle M., Travis M. Gerber, J. Dusty Lane, and Scott A. Ashford. "Lateral resistance of a full-scale pile group in liquefied sand." *Journal of Geotechnical and Geoenvironmental Engineering* 131, no. 1 (2005): 115-125.
56. Seed, R. B., and Harder, L. F., Jr. (1990) "SPT-based analysis of Cyclic Pore Pressure Generation and Undrained Residual Strength," Proceedings of the H. Bolton Seed Memorial Symposium, May, 1990.
57. Schoene, Blair, Kyle M. Samperton, Matthew P. Eddy, Gerta Keller, Thierry Adatte, Samuel A. Bowring, S. F. Khadri, and Bernd Gertsch. "U-Pb Geochronology of the Deccan Traps and Relation to the End-Cretaceous Mass Extinction." *Science (New York, N.Y.)* 347, no. 6218 (2015): 182–184. https://doi.org/10.1126/science.aaa0118.
58. Ashford, Scott A., Ross W. Boulanger, and Scott J. Brandenberg. "PACIFIC EARTHQUAKE ENGINEERING RESEARCH CENTER." (2011).
59. Singh, Kharak, Munukutla Radhakrishna, and A. P. Pant. "Geophysical Structure of Western Offshore Basins of India and Its Implications to the Evolution of the Western Ghats." *Journal of the Geological Society of India* 70, no. 3 (September 2007): 445–458.
60. SNAME - Society of Navel Architects and Marine Engineers. "Recommended Practice for Site Specific Assessment of Mobile Jackup Units, Rev 3, January 2008).
61. Sriram, Kunduri, S. S. Gupte, V. Kothari, Madhu Bisen, and R. S. Waraich. "Structure and evolution of Saurashtra Arch in Kutch-Saurashtra deepwater area, western India." In 6th International conference & exposition on petroleum Geophysics, Kolkata, pp. 21-25. 2006.
62. Stevens, Robert F., and Khell A. Al-Shafei. "The applicability of the Ras Tanajib pile capacity method to long offshore piles." In *Offshore Technology Conference*, pp. OTC-7974. OTC, 1996.
63. Sunarjanto, Djoko, et al. Indonesian Tertiary Sedimentary Basin. *LEMIGAS Scientific Contributions* 31, no. 2 (August 2008): 19–27.
64. Vaidyanadhan, Ramakrishnan, and M. Ramakrishnan. Geology of India. Vol. 2. Bangalore: Geological Society of India, 2010.

65. Van Gorsel, J. T., *Bibliography of Indonesian Geology*, ed. 7.0. July 2018. https://www.vangorselslist.com.
66. Wang, Shin-Tower, and Lymon C. Reese. "Design of pile foundations in liquefied soils." In *Geotechnical Earthquake Engineering and Soil Dynamics III*, pp. 1331-1343. ASCE, 1998.
67. Westgate, Z. J., D. J. White, and M. F. Randolph. "Field observations of as-laid pipeline embedment in carbonate sediments." Géotechnique 62, no. 9 (2012): 787-798.
68. White, David J., and Mark F. Randolph. "Seabed characterisation and models for pipeline-soil interaction." International Journal of Offshore and Polar Engineering 17, no. 03 (2007).
69. Wilson, Daniel W., Ross W. Boulanger, and Bruce L. Kutter. "Observed seismic lateral resistance of liquefying sand." *Journal of Geotechnical and Geoenvironmental Engineering* 126, no. 10 (2000): 898-906.

ABOUT THE AUTHOR

Dr CR Parthasarathy holds BE and ME from Bangalore University, PhD from IISc, Bangalore (1998-2002). He has a Master of Business Law from National Law School of India University (NLSIU), Bangalore in 2025. He is the founder, Chairman & Managing Director of Sarathy Geotech & Engineering services pvt ltd, and established local offices in Indonesia, Vietnam, Abu Dhabi and Saudi Arabia. He is also the founder of Taqell Edification Pvt Ltd., and Chairman of SGES Foundation, a non-profit organization.

Dr. CR Parthasarathy is a distinguished geotechnical engineer and an inspiring leadership mentor whose career spans over three decades. Beginning his professional journey in 1993, he has emerged as a trailblazer in offshore and onland geotechnical engineering, contributing to landmark projects involving multi-storeyed complexes, refineries, bridges, embankments, water-retaining structures, offshore platforms, pipelines, and jack-up rigs across South East Asia, India and the Middle East.

With expertise in site characterization, deep foundation testing and consulting, Dr. Parthasarathy has led and executed numerous large-scale site investigation and engineering studies. As a technical partner of Pile Dynamics, Inc. (USA), he has significantly advanced the quality assurance practices of deep foundation testing in the region. He has achieved EXPERT LEVEL in the PDI/PDCA Proficiency Test on dynamic measurement and analysis. He is also an authorized trainer for high-strain dynamic pile testing, having mentored over 2000 students and professionals in the last two years alone. His commitment to sharing knowledge is evident in the hundreds of technical lectures he has delivered.

He is a Life Fellow of the Indian Geotechnical Society (IGS), past Chairman of the IGS Bengaluru Chapter, and currently serves as an elected National Executive Member of IGS for the third time from 2025-2027. He is the member of Indian National Academy of Engineering (INAE). He also serves as an Executive Committee Member of the Deep Foundations Institute (DFI) – India Chapter and Indian Society of Earth Technology (ISET).

Beyond engineering, Dr. Parthasarathy is deeply passionate about human potential and leadership excellence. He is an Ikigai-certified coach and trained under global legends such as Tony Robbins, Blair Singer, Zoltan Hrotko and others. His unique approach blends technical mastery with personal development, empowering the next generation of engineers and professionals to lead with purpose and impact.

Dr. CR Parthasarathy stands at the intersection of engineering excellence and transformative leadership, shaping the future of civil engineering and professional development across borders.

www.ingramcontent.com/pod-product-compliance
Ingram Content Group UK Ltd.
Pitfield, Milton Keynes, MK11 3LW, UK
UKHW061539051026
15625UKWH00009B/519
* 9 7 8 1 9 5 7 4 5 6 8 3 6 *